RHS

Can Anything STOP SLUGS?

RHS Can Anything Stop Slugs?
Author: Guy Barter
First published in Great Britain in 2018 by Mitchell Beazley, an imprint of Octopus Publishing Group Ltd, Carmelite House, 50 Victoria Embankment, London EC4Y 0DZ
www.octopusbooks.co.uk

An Hachette UK Company
www.hachette.co.uk

Published in association with the Royal Horticultural Society

ISBN: 978 1 78472 4788

A CIP record of this book is available from the British Library
Set in Archer and Open Sans
Printed and bound in China

Mitchell Beazley Publisher: Alison Starling
RHS Publisher: Rae Spencer-Jones
RHS Consultant Editor: Simon Maughan

Conceived, designed and produced by
The Bright Press
Part of the Quarto Group
Ovest House
58 West Street
Brighton
BN1 2RA
England
Publisher: Mark Searle
Associate Publisher: Emma Bastow
Creative Director: James Evans
Senior Editor: Lucy York
Managing Editor: Isheeta Mustafi
Project Editor: Katriona Feinstein
Design: Lindsey Johns
Illustrations on pp43, 71 and 167: Sarah Skeate

The Royal Horticultural Society is the UK's leading gardening charity dedicated to advancing horticulture and promoting good gardening. Its charitable work includes providing expert advice and information, training the next generation of gardeners, creating hands-on opportunities for children to grow plants and conducting research into plants, pests and environmental issues affecting gardeners.

For more information visit www.rhs.org.uk or call 0845 130 4646.

Can Anything STOP SLUGS?

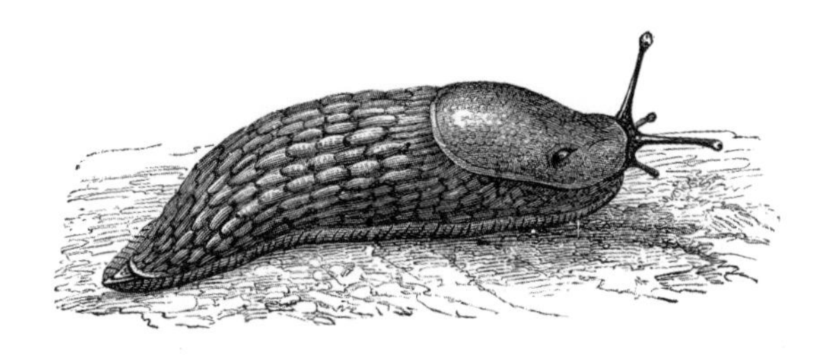

A Gardener's Collection *of* Pesky Problems *and* Surprising Solutions

GUY BARTER

RHS CHIEF HORTICULTURIST

MITCHELL
BEAZLEY

Contents

1 Ornamental Plants

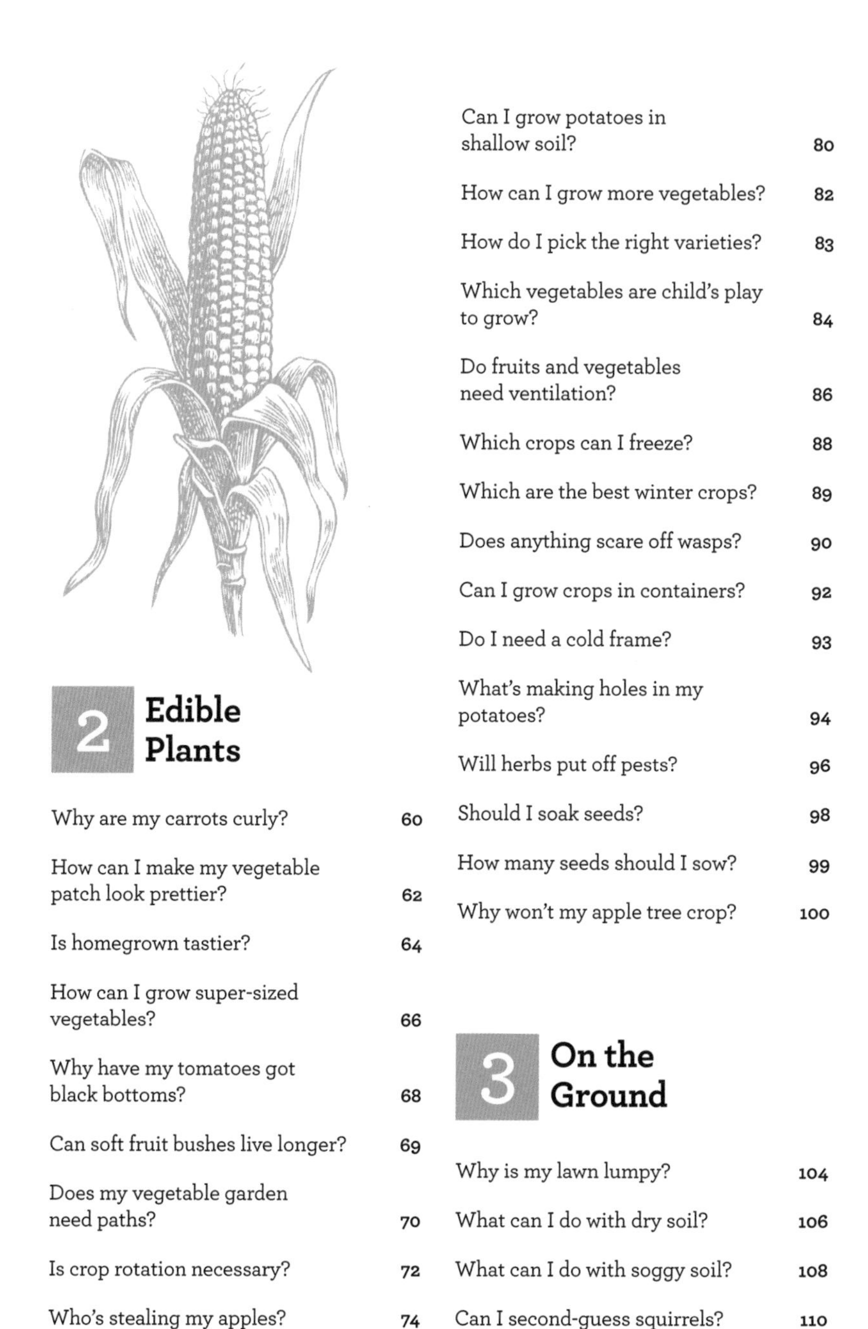

2 Edible Plants

3 On the Ground

4 Everyday Garden Care

5 The Garden and Beyond

Introduction

OF ALL THE QUESTIONS that perplex gardeners, stopping slugs is likely to top the list. It's a question I was asked many a time during my 20 years advising Royal Horticultural Society members and working with the team at the RHS Wisley garden. I have also encountered many less obvious queries over the years. In fact, it would seem there is no end to the things that puzzle gardeners, from the reason why lavender goes leggy to why tomatoes sometimes get black bottoms, and how to stop a lawn from getting lumpy. This book, like its predecessor *How Do Worms Work?*, addresses at least some of these gardening conundrums.

So what about those pesky slugs? Astute readers will realise that if there was an easy way to stop slugs, everyone would have heard of it by now. You can read up on some really helpful practical slug-battling tips on pages 162–163, but the short answer is: it depends. If you only grow trees and shrubs, slugs won't greatly bother you, but if you only grow lettuce and hostas, slugs will almost certainly seriously reduce your gardening satisfaction. But that is typical of gardening: there is no one answer, everything depends. And that is what makes it such an engaging and interesting pursuit. As I hope you will discover in these pages, part of the enjoyment comes from digging into the intricacies of what makes the natural world on your back doorstep tick, and in turn, getting to know your garden will help you to keep things running as smoothly as they can.

Slugs are one of the most irksome of garden pests, but armed with the best tactics to tackle them, you can prevent the worst of the damage to your plants.

About this book

We start by examining some common bugbears about the jewels of the garden: the ornamental plants. You will find out why flowers are so thirsty (the water pressure in their stems keeps them upright, so they need to drink constantly from the soil), why your roses have spotty leaves and how to give your garden instant oomph. Of course, gardening is not just about looks: Chapter Two will help you to get the most out of your edible plants. You'll see which fruits and vegetables are most worth growing yourself to get a better flavour, and how to identify an apple thief (if it's gnawed and left on the grass, it's probably squirrels – if it's left on the tree, suspect birds).

▲ Knowing which plants to choose to attract pollinators will help to improve your garden. Alliums add beautiful splashes of colour and have great appeal for butterflies.

Next we tackle those ground-level problems. What works against weeds? How much should you mulch? And can you make a quick compost? Chapter Four seeks to make everyday garden care more rewarding and enjoyable. Find out how to create your own small-scale wildflower meadow, when you should be anxious about aphids (and when to take matters into your own hands) and how to fill those seasonal garden gaps.

We end our journey by looking at all manner of conundrums about the garden and beyond. Did you know that the flat surfaces around your pond might be putting frogs off, as they can heat up too much to sit on? Or that there are a range of flowering creepers that will happily grow between paving stones, giving even the most concrete of gardens a romantic feel? By the time you finish reading, you'll be armed with invaluable tips and astounding facts, ready to tackle the great outdoors.

A

QUICK ANSWERS

The 'A' box under each question offers you the quick and dirty answer in the shortest form possible. Read on for the main text, which offers additional context and plenty of extra detail.

Chapter 1

Ornamental Plants

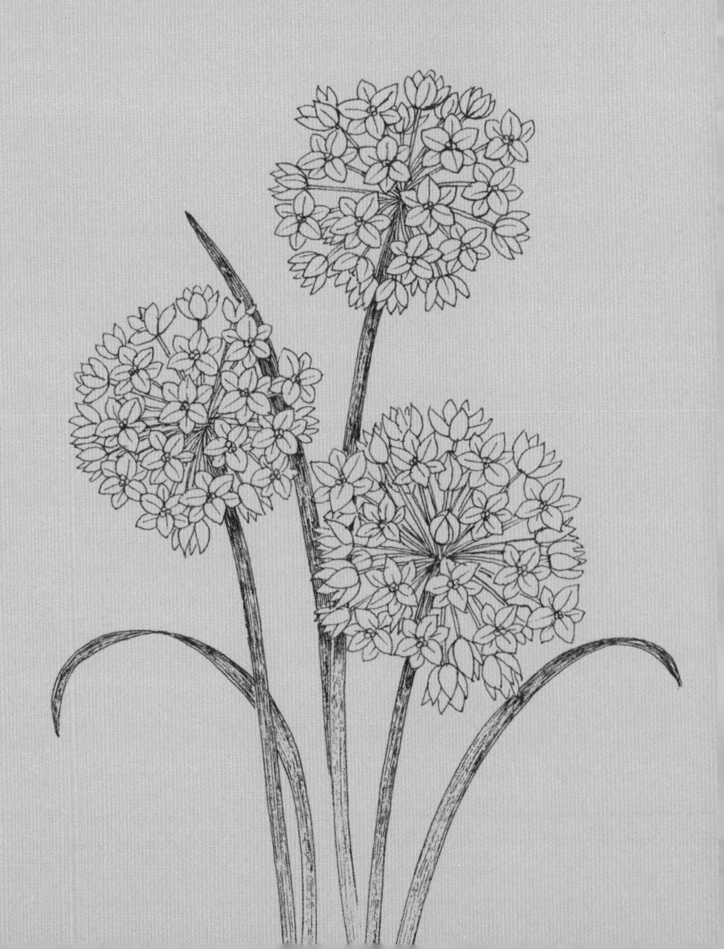

Q Why are flowers so thirsty?

WATERING can sometimes seem like a never-ending chore to the gardener, particularly during sustained periods of warm weather. And since water conservation is often in the news, it seems extravagant, too. Do flowers really need to be watered every day when it's hot?

To get the carbon dioxide they need for photosynthesis, plants must open the pores on their leaves to take in air – and in doing this, they lose water. They use almost all the water they drink to prevent wilting: water is constantly taken in by a plant's roots and passed up to the leaves, and it is the water pressure inside the plant that keeps it standing up. Nevertheless, plants are well adapted and most can cope with a considerable degree of water stress.

Feeling thirsty

Keeping a plant in a pot means that its roots are restricted – they can't travel to seek out more water – and it will need regular watering. Although terracotta pots have the reputation of drying out faster than plastic, this is a relatively minor point: the vast majority of water lost by any pot plants will be through their leaves. If you put naturally thirsty plants such as dahlias and begonias in pots, they'll need watering often – in hot weather, possibly even more than once a day.

Plants that have just been planted also need regular and frequent watering; water will encourage their roots to grow and establish themselves. After around three weeks most plants will have settled in, and their roots will be in good enough shape to go forth and find water. Established plants generally only need watering during prolonged dry spells, in which case a soaking every 10 to 14 days will be enough.

Tactical watering

When you do water, make sure that water reaches the roots of your plants. In summer, water applied from above will fall on the leaves but may not get down into the soil: you need to add enough water, slowly enough, to make sure that it reaches the roots. In very hot weather, it's worth watering in the morning and allowing enough time for the water to sink in evenly across the

A Although plants need both water and food to thrive, there are only a few circumstances in which they need frequent watering. Plants grown in containers and those that are newly planted usually do need watering regularly and often.

whole soil area – particularly if you're watering on uneven ground, where a quick soaking may mean that some areas flood while others are left dry.

When a garden is large and demanding or when its owners are often away, watering systems can help out. Low-level seeping or dripping systems can be left on a timer and are a good way to ensure that established plants are watered consistently, even if no one is around. Sprinkler systems are wasteful in comparison, but may work better for newly planted areas, in which case consider mini-sprinklers rather than the full-sized variety.

PLANTS THAT DON'T NEED MUCH WATER

Stonecrops, *Sedum*. The thick, succulent leaves of sedums act as their own water store, enabling them to sit out dry periods without damage. The ice plant, *Hylotelephium spectabile*, is a close relative of the sedums and also very adaptable.

Ice plant, *Hylotelephium spectabile*

Lamb's ear, *Stachys byzantina*. This is probably the most familiar form of *Stachys*. It is a useful, attractive evergreen perennial, with grey, furry leaves and pinkish-purple flowers growing in whorls up the stems.

Lavender, *Lavandula*. Lavender plants have the needle-like leaves of many drought-resistant plants and there's a huge range to choose from, from the traditional violet-blues to whites and pinks.

Eryngos, *Eryngium*. Prickly customers with teasel-like flower heads, eryngiums come in annual, biennial or perennial forms, in a range of silver, blue or blue-green tones.

Russian sage, *Perovskia* 'Blue Spire'. This is a beautiful deciduous plant with tall spires of violet-blue flowers. It has a strong fragrance somewhere between sage and lavender.

Alpine eryngo, *Eryngium alpinum*

How much water is too much?

How can you tell when you've watered your plants enough to make a difference? Also, I've heard that overwatering my pots may cause more problems than giving them too little water – is this true?

More water than is needed to fill the gaps in the soil around plants' roots is water wasted. The surplus will either drain away or, if drainage is really slow, will displace the air in the soil – which can result in suffocated roots.

Plants need to breathe through their roots, as well as to absorb water. If the soil gets waterlogged, the water will force the air out and the roots will not be able to get the oxygen they need.

Overwatering in containers

While plants in soil aren't usually killed by overwatering, it's comparatively easy to overwater those in containers, particularly if you're a conscientious waterer living somewhere where it rains a lot. Planting in a porous potting medium that will drain easily will help but if your container plants have yellowing leaves, it's a sign that you are probably overwatering. Check your containers by gently tipping the plant out – if the roots are dark, don't have fine hairs on them, and look brittle, they are starting to rot. There may also be a slightly sour smell around the roots and the soil, too.

Soil drainage

If you garden on heavy, clay-rich soil that absorbs water slowly and releases it even more slowly, raised beds are an excellent solution – although they take a little time to set up. The no-dig system (see p71) can work well, provided you have paths to ensure you don't compact the soil by walking on it when it's wet: undug but mulched soil encourages a high worm population which will aerate the soil for you.

Earthenware pots lose moisture through their porous surfaces. This helps to reduce the risk of overwatering.

Why won't my 'bulb lasagne' work?

THE IDEA OF PLANTING UP A CONTAINER of bulbs that will flower in succession may sound appealing, but in reality it can prove a little more complicated. Some pots aren't deep enough to take the suggested layers and messy foliage can interfere with later-flowering bulbs.

In a 'bulb lasagne', different flower bulbs are planted in layers, so that they flower in sequence over an extended period. It's an idea that features regularly in bulb catalogues and gardening magazines. Although some of those perfect pots will have been mocked up to create beautiful photographs, there's no need for that to put you off: you can get good results by following a few rules.

Most people start with just two layers – perhaps crocus, say, followed by tulips. The lowest layer should be the last to flower, so in that case, the tulips would be planted first, covered with 5cm of compost, followed by the crocus, topped off with another layer of compost.

Ideally bulbs should be planted at three times their height, although this may not always be possible unless the container you use is very large. Work out the relative depths before you start, and if the final layer of bulbs ends up very close to the surface, cover the soil with an upturned pot to protect them for some weeks after planting. Keep the density of bulbs a little less than you would if planting in a single layer – you need to leave enough room for the lower layer or layers to grow through the top one, so allow a two-finger width space between the bulbs in each layer.

Consider foliage as well as flowers when you're choosing your bulbs – hyacinths, narcissi and tulips usually have large, quite coarse leaves, which means that they do best at the end of the sequence if you're also using small, finer-leaved bulbs such as dwarf iris, crocus or grape hyacinths.

Layering bulbs is a simple principle, and is a good way of mixing colours and giving a planted container a long flowering life. Only a few things can go wrong: you might have planted the bulbs in the wrong order, meaning the results aren't as pleasing (although they'll still flower), or planted them too deep or conversely, not deep enough.

Q How can I get plants for free?

IT'S TEMPTING to plan your ideal garden from the most appealing plant catalogues and nursery displays, but plants can be expensive, and with even a smallish plot funds may run short very quickly. What are the best ways to source free plants?

A There are numerous ways to get plants for free – but be discerning: just because it's free doesn't necessarily mean that it will suit your space. And beware of friends bearing gifts: they could be offering you weed-like plants that have become invasive in their own gardens.

Start by checking if there are extra plants in your own garden – you'll often find seedlings of plants such as hellebores close to their parents, and they can be nurtured and moved elsewhere. You can also collect seeds from existing plants, then, depending on the species, either store them somewhere cool and dry to sow in the spring, or sow fresh into the ground (tougher seeds with thicker coats often need a period of winter chill in order to germinate in the spring). Look, too, at plants that have formed large clumps that lend themselves to dividing (see box) and learn to take your own cuttings (see pp48–49).

Looking further afield

Friends and family will often donate seeds, cuttings or seedlings: if they've over-sown, they'll have extras. Visit plant sales, too – even thrift sales often have plant offerings, and if you're in the country in spring, you'll also see 'honesty tables' at the roadside offering cheap or even free plants.

A number of plant societies such as the RHS, the Hardy Plants Society or the Heritage Seed Library make distributions of seeds to their members, which can be a good and economical way of getting interesting new species.

A few precautions

So-called 'finger blight' is looked on poorly by respectful gardeners. It is bad etiquette to help yourself to cuttings from someone else's garden, public or private, without permission. Try asking – the gardener will often spare you a cutting or two if they can.

Take care what you let into your garden if you don't recognise it or know its background. Seed is a safe source, but you can unwittingly let pests and diseases in with new plants – separate out arrivals for two or three weeks until you've seen they're healthy, and identify

GOOD DIVIDERS

Many plants form clumps that are easy to split and divide when they're dormant. Five examples that divide well:

- **Plantain lilies, *Hosta***
- **Heucheras, *Heuchera***
- **Primulas, *Primula***
- **Montbretias, *Crocosmia***
- **Herbaceous peonies, *Paeonia***

How you divide a clump depends partly on the habit of the plant you're dividing and partly on the size of the clump you're tackling. Smaller clumps of plants with fibrous roots, such as heucheras and hostas, can be gently pulled apart by hand.

Crocosmias form either corms or fleshy underground stems called stolons, which can be pulled apart, each with its own root. A large, firmly rooted clump of herbaceous peony, for example, may need to be cut to divide it. Any remaining foliage should be removed, then the clump eased up as gently as possible. The earth should be washed off so roots and dormant buds can be seen clearly. Sections can be cut out with a clean, sharp knife, each with a minimum of three buds, and then replanted with the buds 5cm below the surface of the soil.

Plantain lily, *Hosta*

unknowns before planting them out – they may be thugs. For the same reason, treat gifted plants with caution – a free goldenrod (*Solidago*) plant or two, for example, might look appealing, but the family is famously invasive.

Keep a garden plan going, and make sure new finds or gifts have a place in the garden. Free is always good, but you don't want to end up with too many mismatched plants that don't work well together.

Why are my containers so underwhelming?

YOU HAVE MORE CONTROL over the different aspects of planting in containers than you do in most kinds of gardening, so it's especially disappointing when the results fail to live up to expectations. Look at the ways they fall short – have the plants failed to thrive? Is there plenty of growth but no flowers? Were the flowers shorter-lived than expected? Or, if the plants grew well, did the overall look have less appeal than anticipated? Then seek out the specific solution.

LOOKING GOOD

If containers lack visual impact, it may simply be because the plants in them are too small. If you want a plant to make an impression but don't want to wait a long time for it to grow, you may need to buy a bigger one in the first place. Dahlias, for example, can reach an impressive size in a couple of months and will certainly offer plenty of visual oomph in full flower, but will need lifting, storing and replanting annually. Alternatively, you could consider seasonal or long-term planting. When carefully looked after, trees, such as Japanese maples, will make a graceful impression for years.

The grouping of containers is also important: a single tall pot planted up with a mix of flower and grass – a colourful *Geum* and a *Pennisetum* for example – might look stunning, but so might a tabletop of small containers planted with intricate tiny succulents.

A collection of small-scale cacti and other succulents makes more impact in a group than they each would individually.

A

All the possible answers fit into one of two camps: either the plants in the containers didn't do well, or, if they did what they were expected to do, somehow the effect wasn't as good as you thought it would be. Look first at the health of the plants, then at the overall aesthetic effect.

If your plants didn't grow much, it could be as a result of a lack of light, not enough – or too much – water, or not enough fertiliser. If they grew well but failed to flower, this could also be due to a lack of light, or too much nitrogen-rich fertiliser – which prompts leafy growth – in the potting mix.

In some cases, if the plant is short of water at the point when the flowers are developing, it will fail to flower completely. Camellias are a case in point here; if they are short of water in the late summer, you will find that their flower buds don't set.

When plants flower but the flowers are very short-lived, or if the leaves die back early, it may be as a result of too much sun, or lack of enough water – drought stress.

In almost every circumstance, it's also possible that the plant or plants you bought weren't of the best quality, in which case, all you can do is research your chosen cultivar, and choose as carefully as you can next time.

Getting it right

You can grow most kinds of plants in containers, provided that you cater for their likes and dislikes. General points include always making sure the container is the right size for the plant or plants: it needs to be large enough to accommodate the roots and to allow for some root growth, but not so large that a small plant is sitting in a sea of compost. It should have enough holes in the base to allow it to drain freely. Standing pots on blocks can help ensure that they don't become waterlogged.

Container plants can be fed through the summer with a liquid fertiliser, starting between four and six weeks after planting up. In hot weather most will need watering at least once a day; take the opportunity to check for signs of slug or snail damage, or of any other pests and diseases, and remove any dead leaves and flowers. Daily deadheading is a good habit to get into, as it helps prolong a plant's flowering season.

▼ A group of potted geraniums looks effective in a larger container, such as this wooden crate.

Why don't my roses smell?

THE CLASSIC ROSE SCENT is so appealing that for many gardeners it's almost as important as the beauty of the flowers when it comes to choosing which variety to grow. But roses are not all equal when it comes to fragrance, and different factors play a part in how strongly they smell.

In 2015 a group of French scientists at the University of Lyon discovered that one enzyme was far more active in the petals of a strongly scented rose than in a rose bred for other qualities. Named RhNUDX1, the enzyme works in the petal cells, helping to generate a substance called monoterpene geraniol, which is a key ingredient of rose oil. The discovery seems to indicate that, given time, it may prove possible to reintroduce and boost scent in roses that, over years of breeding, have lost their smell.

Breeding roses

Cultivated roses were first introduced to Europe from China towards the end of the 18th century, and breeders have been experimenting ever since. Today, roses divide into three main groups: species roses, old garden roses and so-called 'modern' roses, creating a huge range that includes shrubs, climbers, ramblers and small and miniaturised roses, in almost every colour (including a slightly unsettling lilac-blue). Objectively, some types have hardly any scent, having been successively bred for other qualities, such as repeat flowering, disease resistance and particular strong colours, while others (mostly falling in the 'old' garden roses category) still smell powerfully of rose. To find your ideal scented rose, follow the rose breeders' descriptions online or in catalogues, or, best of all, search them out when they're flowering so you can tell when one is your perfect match.

A: Scent is subjective, of course, and intensely personal – some people seem to be more sensitive to rose scents than others. Nevertheless, some roses are notably more powerfully scented than others, and warm sun brings out the scent most strongly. There are literally thousands of options to choose from – the American Rose Society lists over 15,000 – so you should be able to find the right one, with the right scent, for you.

What roses like

Check you have the right situation for a rose before you buy. Although the majority are sun-lovers, some varieties will tolerate shade. Ensure you know what size a new rose will grow to – some, especially the rambling type, can grow vast, while at the other end of the scale are a number of small roses for little gardens. Some roses will manage in containers, too, although these aren't usually their ideal habitat. If you do want to grow a rose in a container, choose the variety carefully, and plant it in a soil-based potting medium. Breeders have had a high degree of success with disease-resistant varieties, though roses can still suffer from a number of pests and infections (see pp30–31 and pp142–143). If a rose succumbs, avoid planting its successor nearby.

SIX ROSES WITH STANDOUT FRAGRANCE

'Gertrude Jekyll'. English shrub rose, which will also climb, named for the celebrated gardener, with pink flowers and a classic rose scent.

'Golden Celebration'. A classic English shrub rose, with, as its name suggests, large, rich golden flowers and what David Austin, its breeder, describes as a 'delicious rich tea scent'. Can be grown in containers.

'Madame Isaac Péreire'. Very strongly scented, deep madder-pink flowers with gold centres, an old shrub rose, which will also climb.

'Lady Emma Hamilton'. Available in shrub or standard forms, orange-gold flowers with a fruity scent.

'Guinée'. A climber developed in the 1930s, with very dark, velvety red flowers. Not the easiest to grow, but with a powerful deep scent.

'Scarborough Fair'. A small shrub rose with soft pink, open, cup-shaped flowers with gold centres and a musky scent. Suitable for smaller spaces; only grows to a height of 75cm.

Q Why has my wisteria never flowered?

WITH ITS TRAILING TASSELS of subtly coloured flowers, wisteria has such a knockout appearance when it's grown successfully that it appears on a lot of people's wish lists. But while it's generally not too fussy, it can be slow – sometimes excruciatingly slow – to flower, and it can also be given to sudden, catastrophic die-back.

A There are a number of possible reasons for a wisteria to be sluggish when it comes to flowering. It may be slow to mature, and often takes its time after planting to settle in (wisterias grown from seed can take up to 20 years to flower). Insensitive pruning may also be a factor, or it may not be sited in a spot it likes.

If your wisteria was planted in the last year or two, you may be being impatient. The roots can take some time to establish themselves in a new spot; give it another year or even two and your patience will probably be rewarded. This assumes that you bought a named variety from grafted rootstock, which is not only faster to flower but also more reliable than wisteria grown from seed or cuttings – and it's also best to buy a wisteria that is already in flower, demonstrating that it's active.

What to do

If a wisteria has been planted for several years and appears healthy but still isn't flowering, there are other possible factors that may be discouraging it. Make sure that the soil around its roots contains plenty of well-rotted organic matter. If the soil is naturally poor, it may mean that the wisteria is low in potassium, in which case a spring feed of sulphate of potash, using 20g per square metre, will help.

Although wisteria flowers in spring, the flower buds begin to develop from the summer of the previous year, and it's important to make sure that the plant is kept well watered between midsummer and autumn – if it is allowed to dry out during this period, the buds will stop growing. An unlucky spring frost can cause unavoidable damage – if it is sharp enough – making the wisteria drop its buds before it flowers.

Wisteria is usually pruned twice a year – in high summer, after it has flowered, to take the green shoots back to where the buds will develop, and in early spring when it is cut back to where the new buds are already visible. Careless pruning can take the flower buds as well as the shoots, so it's important to know what you're doing.

Problems

Wisteria isn't particularly problem-prone, although it can suffer from infestations of scale insects or attacks of coral spot (which causes branch die-back) or powdery mildew on the foliage. But it has one specific peculiarity, which is that the original graft may suddenly fail, even in fully mature and apparently healthy plants. This means that the graft on which the original rootstock was made separates, and it will cause most or all of the top growth of the wisteria to collapse and die. The only way you can try to guard against it is to check that the join is clean and firm when you first buy the plant.

SITING A WISTERIA

Once planted, they don't like being moved, so it's worth ensuring that you have the right site for a wisteria before you put one in. The classic situation is a sheltered location growing against a warm wall, but if that's not something you can offer, it can also be allowed to scramble through a tree (in which case it can be left, unpruned, to its own devices), trained across a pergola, or trimmed into a standard, small-tree shape – in the latter form, rather less familiar today, it was a popular addition to Victorian parks; modern gardeners will sometimes grow one in a container.

When wisteria is planted in a favourable situation, it can be extremely long-lived. Plants over a century old are not uncommon.

Q Do I have space for a cutting garden?

A CUTTING GARDEN is one grown specifically to supply cut flowers. Given this, it is planned more like a vegetable plot – with an eye to maximum productivity rather than to artful design – although a well-planned cutting garden is pleasing in the same way as a healthy vegetable plot. Can a modestly sized garden expect to support one, as well as conventional flower beds?

A

If you have space for a largeish dedicated bed – of a few square metres – you have enough room for a cutting garden. Well-chosen flowers can be immensely productive, so this sort of area will more than suffice to give you flowers for the house almost year-round.

Dahlias are great double-duty plants: the flowers are both striking in the garden and long-lasting in the vase.

If you do have space for a cutting garden, make it part of your overall garden layout in the same way as you would a vegetable bed. If you don't have space, think about incorporating flowers – both annuals and perennials – that are prolific enough to 'cut and come again', so that you can cut flowers for indoors while leaving plenty in the bed. A wigwam of sweet peas, for example, or a strong dahlia, will flower prolifically across many weeks – and the more flowers you cut, the more will grow. Zinnias, cosmos and multi-stemmed sunflowers are other generous croppers that will yield masses of flowers.

Adding greenery

Don't forget greenery – even flower-heavy arrangements need a green backdrop as a foil. Evergreen hedges offer useful sources for this: viburnums, oleasters, eucalyptuses, hollies and ivies all make good additions, as do sprigs of herbs such as rosemary, thyme and lavender.

A YEAR'S WORTH OF CUT FLOWERS

Plan ahead to make sure that your cut-flower crop will offer something new every month between April and October, with fresh flowers following on as others end their flowering season:

- April — Daffodils (*Narcissus*), tulips (*Tulipa*)
- May — Bearded iris (*Iris*), late tulips (*Tulipa*)
- June — Larkspurs (*Consolida*), bachelor's buttons (*Centaurea cyanus*), delphiniums (*Delphinium*)
- July — Sweet peas (*Lathyrus odoratus*), lilies (*Lilium*)
- August — Sunflowers (*Helianthus* – particularly smaller multi-stemmed varieties), gladioli (*Gladiolus*)
- September — Dahlias (*Dahlia*)
- October — Asters (*Aster*)

Bachelor's buttons, *Centaurea cyanus*

Hellebores will give you flowers to cut in deep winter, too. Upright-flowering varieties are now available, but if your flowers grow face-downwards, you can still make an arrangement by floating the flowerheads in a bowl of water.

Keeping cut flowers in good condition

Pick flowers in the morning, choosing those that are still in bud, rather than in full flower – they'll last longer in the vase. Strip the lower leaves from each stem (if they're submerged in water, they'll rot and smell bad) . Cut the stems at an angle with a sharp knife or scissors and leave them in a cool place in a bucket of water for a couple of hours before arranging in your container. Flowers with very flexible stems such as tulips may bend and flop when arranged – to prevent this, roll the stems in few sheets of newspaper when leaving them to soak before arranging. Dissolve some cut-flower food in the water in the container before you add and arrange the flowers.

Do I have time for a flower garden?

A LOW-MAINTENANCE flower garden is within the reach of most gardeners so long as they choose carefully. And the choice is wide, helped by the fact that in recent years plant breeders have concentrated on developing plants that are both easy to care for and offer a long flowering season.

Coneflowers, *Echinacea purpurea*, are tough, popular perennials that enjoy full sun. In the right conditions, they can flower for years.

Aim to plan in advance so that the planting can be done in one go, rather than buying individual plants and slotting them in more randomly. The best strategy for a low-maintenance garden is a careful choice of base plants. Read descriptions in detail before you buy, bearing in mind how much sun or shade your garden has, and what type of soil, and consider what the plant can offer you. When you plan, think of perennials and perhaps some of the smaller decorative grasses as long-term, then you can fill in the gaps with spring and autumn bulbs and some easy annuals.

> If you have time to plan a low-maintenance flower garden, you probably have enough time to keep it going. Begin by planting herbaceous perennials, which are low-care and will return year after year, then fill in the gaps with easy-to-grow annuals.

Keeping things tidy

When plants are in place, the task that takes the most time – even in a low-maintenance garden – is weeding. Little and often is the best rule to keep weeds under control; if you leave it too long, it becomes a much larger job. Quite dense planting will help; if there isn't a lot of bare soil, there isn't much space left for weeds. If time is limited, tidying the edges of beds and – if you have them – weeding paths and paving instantly makes a garden look tidier and more under control.

CHOOSING PLANTS THAT LOOK GOOD TOGETHER

To get the best visual effect out of low-maintenance plants, it can be simplest to think in terms of pairings that will look good together. You can pair like with like (perennials with perennials and annuals with annuals), or put annuals along with perennials to try out different effects in successive years without having to move or change perennials.

A few ideas:

The bright, open flowers of the annual cosmos, whether you go for the white, pink or red varieties, look good growing amongst the silvery leaves of one of the smaller *Miscanthus* grasses, while the vivid blue of a perennial geranium (Rozanne 'Gerwat' is an enduringly popular one) will 'pop' with the bright orange of calendula, the common marigold, which has a long flowering season.

Look at achilleas and sedums, both perennials available in a range of colours and heights, which form good-looking clumps and provide a structural 'background'. Euphorbias, also perennial, come in a wide spectrum, too, most green, but with acid yellow (*E. palustris*) at one end and vivid red ('Dixter') at the other – the acid green or yellow ones look striking mixed with purple alliums, both at their best in late spring. Perennial wallflowers mixed in with tulips are another classic and easy-to-grow spring combination.

The brilliantly green, dark-eyed Martin's spurge, *Euphorbia* x *martini*, makes a striking contrast with the purple of *Allium atropurpureum*.

Q Why are my lupin roots lumpy?

SOMETIMES lupin roots have pea-sized bumps on them that look like growths. Does this mean that there's something wrong with them – are they diseased, or is there another explanation?

Lupins are a member of the Fabaceae, or bean family, usually called legumes; with over 19,000 members, it's one of the largest plant families. The rhizobium bacterium is commonly found in the soil and, when it enters the roots of a legume, the two form a mutually beneficial relationship: the host plant's roots give rhizobia a home, while the bacteria help their host to access the nitrogen that it needs for growth. The relationship is a great example of complex natural chemistry. Inside the nodule, the legume prompts gene action in the rhizobium that enables the bacterium to convert nitrogen gas (which the plant can't use directly) to nitrogen compounds (which it can). Before the age of commercial fertilisers, legumes' production of nitrogen and the benefits it brought as a natural fertiliser were very valuable indeed to agriculture. The soil-enriching qualities of the Fabaceae family was noticed in ancient times: Cato, the famous Roman nature writer and agriculturalist, wrote about it as early as the 2nd century BCE.

THE IMPORTANCE OF ROOTS

Gardeners are used to studying plants' above-ground structures as a guide to their overall health, but tend to overlook the state of their roots. This is a mistake – roots offer valuable clues to potential problems with a plant's well-being, and often, by the time the leaves and stems are showing signs of stress, it's too late to do much to save the plant. If a plant dies on your watch, get into the habit of digging up the casualty and conducting a post mortem on the roots: it may reveal previously unsuspected pests or diseases, or cultural problems such as poor drainage.

A The lumps are actually useful to the plant. They form when a specific bacterium, rhizobium, grows around the natural root hairs, and it's a symbiotic relationship.

Q Why are some plants sold with bare roots?

THE MAJORITY OF PLANTS you can buy in a nursery (and many that are bought online, too) come in containers, planted in soil. So why, when you order some plants, such as roses or fruit trees, do they arrive bare-rooted, without any soil to support them? Don't they suffer without soil?

A For those that can deal with it, it is easier and lighter – and therefore cheaper – to transport plants over a long distance with bare roots. Environmentally speaking, it's also economical, as it dispenses with the use of subsequently unwanted pots, or ecologically unfriendly peat-based compost.

Buying plants with bare roots does mean that the gardener thinks through their purchase in advance; it's a good antidote to those impulse buys that are sometimes made without any real thought as to the place they will have in the garden. Plants that are sold with bare roots commonly include roses, many shrubs and fruit trees and bushes – and the quality of the stock is often better than that of pot-grown ones.

What to do when they arrive

Bare-root plants are transported in winter, when they're in a dormant state, so may arrive looking rather like dead sticks. Ideally they should be planted immediately, but if for any reason this isn't possible, they should be kept somewhere cool but frost-free, such as a shed, to ensure that they remain dormant. Cover the roots with damp straw, or potting compost in a plastic sheet, to make sure that they don't dry out.

Alternatively, they can be planted temporarily in a trench, ensuring that the roots are covered with soil to a depth of at least 20cm, firmed down gently and watered – this is called heeling in, and means they then can be safely left until you are ready to plant them in their permanent site.

Bare-root roses are both easier and cheaper to transport than those supplied already potted in containers.

How can I stop my rose getting spotty leaves?

In early summer, roses can break out in orange acne, as the top of the leaves develop spots and pustules (spore masses) appear on the leaves' undersides. Known as 'rose rust', this is unsightly in the short term and weakens the plant in the long term, so it is worth remedying.

As always, good plant husbandry is a great starting point. Ensure the plants never want for water and get a good mulch of organic matter at least once a year (though this should never touch the stem). Prune the plants correctly and promptly every winter, then remove and preferably burn all fallen leaves and pruned stems.

> It might be an old cliché, but 'prevention is the best cure' is the answer here. Fungicidal sprays should be a last resort.

There should be good air circulation around the plant. A humid atmosphere can foster fungal diseases such as rust, so consider removing or pruning other plants around the rose if they are too close.

In spring, regularly check the new growth for signs of infection, and remove any affected stems by cutting back into healthy growth, burning the prunings to destroy the fungus. Unchecked, rust will develop brown winter spores to replace the summer orange ones, and when the leaves drop for winter, you will need to clear them all away to ensure the spores don't reinfect the plant in spring.

Rose rust can also cause stems to distort and crack, making the plant more vulnerable to secondary infections. If all else fails a fungicidal spray can be used, following the application instructions carefully.

Don't let rose rust (pictured), mildew or black spot spoil your beautiful blooms. Good gardening practice will help you prevent fungal problems of this kind.

FIVE OTHER COMMON ROSE PROBLEMS

Many rose cultivars have been bred to be resistant to rust (which is caused by the fungus *Phragmidium*), but there are several other rose rogues to watch out for:

Rose black spot, *Diplocarpon rosae*

- Black spot is also caused by a fungus (*Diplocarpon rosae*) and can be treated with a suitable fungicide.
- *Podosphaera pannosa* is another fungus, which causes powdery mildew (see pp142–143). As the name suggests, this results in a white, powdery growth on the leaves and shoots.
- Leaves can end up looking 'burnt' (yellowing, browning and withered) if you use too much fertiliser on the soil.
- Yellowing leaves could also signal a lack of water or nutrients, such as nitrogen, magnesium and iron.
- Young stems can be distorted by aphid infestation, which can also turn the leaves black as sooty mould develops on their 'honeydew' excretions.

Why are roses grown in vineyards?

Roses planted next to rows of vines are not there to look pretty; they serve an important purpose. As the rose falls foul of mildew earlier than the vines do, they can give an advance warning of impending infection. They also provide winter habitats for beneficial insects.

Much loved by generations of gardeners, roses have practical benefits as well as stunning looks.

Q What makes my bulbs 'blind'?

A GROUP OF DAFFODILS that has flowered well for the last two years comes up with healthy, green foliage – but no flowers. Why do bulbs go 'blind' in this way, and can the problem be reversed?

A

There are a multitude of possible causes for this, and while there are various fixes you can try, there's no guaranteed cure.

The first thing to do is to try and work out the root cause of the bulbs going 'blind'.

If the daffodils were growing on a dry site, the bulbs may not have got the nutrients they needed after their last flowering. This means that they couldn't bulk up sufficiently to flower in the following year. Cutting the leaves down too soon after flowering (or, in the now rather old-fashioned practice, knotting them for a neat effect) can lead to the same result.

If the bulbs were originally planted at too shallow a depth, it may have encouraged them to divide, creating tiny bulblets that don't have enough resources to make flowers. It's also possible that the plants grew and reproduced too successfully, and things are now getting a bit crowded underground, meaning that each individual plant has less access to the nutrients it needs.

An alternative cause is that the bulbs may have been attacked by pests – narcissus fly, in particular. Dig up a bulb or two and look for symptoms: if the centre of the bulb is rotted and filled with a muddy substance, or sometimes a single maggot, large narcissus bulb fly is responsible. This irritating pest can also attack other members of the Amaryllidaceae family,

A dense mass of 'blind' daffodil foliage above ground may mean that the bulbs have become overcrowded under the soil.

FIXES TO TRY

- Make sure the bulbs are planted deeply enough. Remember when planting that, in the wild, daffodil bulbs grow from depths as great as 60cm. Loosen the soil with a fork to get the bulb planted in, then firm it down afterwards (this can also help to prevent flies getting down to the bulbs to lay eggs).
- If there's no rain in the weeks after the bulbs have flowered, water them – this is the period when the plants need to build up their resources for next year. If the bulbs are in a container, feed them with a solution of tomato feed; if they're in open ground, scatter the soil with a generous sprinkling of slow-release fertiliser granules.
- Let the leaves stay undisturbed on the plant for at least six weeks after flowering, or better still, until they begin to yellow and die back naturally.
- If you suspect the bulbs may be getting crowded, dig them up after they have flowered but before they die right back, and space them out to give them more room.

Daffodil foliage should be left to die down naturally after flowering is complete; it benefits the bulb and creates energy for future flowering.

including snowdrops and nerines. If you find several smaller maggots, the culprit may be small bulb flies, which are types of hoverfly (*Eumerus strigatus* and *Eumerus funeralis*). Affected bulbs should be taken up and destroyed to avoid the problem spreading. Blindness can afflict other types of bulbs, too. Some bulbs, in particular tulips and hyacinths, seldom grow back 'true' after the first year's flowering, and many gardeners now routinely replace them every year to fix this problem.

Q Can I curb my climbers?

If a large climber is getting out of control, is it safe to prune it back hard? And how much art is involved in pruning – how difficult is it for an amateur to undertake?

This may be advice after the event – but it's clearly a good idea to know how large your climber has the potential to grow before you plant it. Current plant breeders have plenty to offer if you're looking for a climber or climbers for a comparatively small space – they have bred many types that are content in containers and are small enough to go up a patio wall without overwhelming their surroundings.

Of course it may be that the climber is perfectly suited to its surroundings, but still needs cutting back. Some vigorous types can become extremely large – the 'Rambling Rector' rose, for example, or *Clematis montana* – but they will also tolerate being cut back quite hard, provided you do it at the right time of year. For the rose, that will be in winter, and in the case of the *Clematis montana*, it should be done after flowering.

Mountain clematis, *Clematis montana*

A

Many vigorous climbers can tolerate hard pruning, but it's important to know what time of year to prune, and how far you can go.

If you have an outsize climber, for example, one that requires a ladder to prune, make sure you have the right tools and knowledge before tackling it. If in doubt, call in a professional.

GET YOUR TIMING RIGHT

The rule of thumb for pruning climbers is that a plant that flowers on the previous season's growth should be pruned immediately after it has flowered, which will generally be between winter and early summer. A plant that flowers on the current season's growth – usually between midsummer and autumn – should be pruned in late winter or early spring.

Why do my flowers' Latin names keep changing?

THE IMPORTANCE OF USING an international language to identify plants is clear. But why do the Latin names of flowers keep changing? Why, for example, a few years ago, did one of the common Michaelmas daisies, *Aster novi-belgii* (right), become the harder-to-remember *Symphyotrichum novi-belgii*?

To make things even more confusing, name changes can belong in two categories: taxonomic and nomenclatural. A nomenclatural change comes from the international body that fixes plant names, and simply dictates that, internationally, a plant will go by one name rather than another. A taxonomic change is more complicated; this relates to the actual classification of the plant. And the more that is understood about plant genetics, the more that is uncovered about the origins of and relationships between plants, which may lead to them being classified under different groups. Taxonomic changes are proposed by botanists. Whether or not a change is accepted and passes into common use depends, in the UK, on the Nomenclature and Taxonomy Advisory Group, a number of experts on plant classification. They will take all the evidence into consideration, consider the opinions of botanists all over the world and decide whether or not the taxonomic change should be made. If it is, the new name will appear in the next edition of the *RHS Plant Finder*, the plant bible of the UK.

And why the change in the Latin name of Michaelmas daisies? '*Aster*' was originally coined by Carl Linnaeus, the inventor of the Latin plant-naming system, but recently these plants' close relationship to native North American asters was confirmed, and as a result they were put into a smaller, more specific genus. So the simple *Aster* became *Symphyotrichum*.

The answer lies jointly in history and science. When the historical names were chosen they related to contemporary knowledge of plants' genetic inheritance, some of it dating back over two centuries. As new discoveries are made, they sometimes change the picture of where groups of plants best belong.

Q What works well with grasses?

DECORATIVE GRASSES are as popular as they have ever been in the domestic garden and the huge range available means that you can find something that works well in almost any spot. What are the best growing partners for them in the garden?

Grasses are both fashionable and flexible: there's a huge range available, in almost every size, including evergreen and deciduous, and sun- and shade-lovers. They don't usually cause many problems for the gardener unless they are in the wrong spot, though many will need regular pruning and dividing in spring if their clumps get too large.

Whether you mix them with their own kind or in combinations with

GRASS, SEDGE OR RUSH?

A plant that looks grass-like may not necessarily be a grass. Two other families, the sedges and rushes, share some grass characteristics, although an old rhyme identifies the key difference:

'Sedges have edges, rushes are round, and grasses are hollow right up from the ground.'

Despite their differences, sedges and rushes are often casually grouped together with the true grasses under the overall heading of 'ornamental grasses'. Sedges are actually part of the Cyperaceae family, with stems that are often triangular, rather than round, and full of spongy pith, and rushes are Juncaceae. Grasses belong to the Poaceae family. They're usually hollow-stemmed and often have stems with solid, swollen joints or nodes.

However, it's not nitpicking to differentiate between the groups; it can affect the gardener's decision on the best place to site them. Although grasses tend to thrive in sun on well-drained soil, sedges and rushes usually like moister soil conditions, and some will also do well in shade.

other sorts of plants, make sure they have company: there are few things more forlorn than a lone 'specimen' grass. The main key to partnering them, after you've checked matching cultivation requirements, is to get the size pairings right, so that neither partner dwarfs the other. Choose perennials that continue to add interest even after they have flowered – such as *Aster, Rudbeckia, Echinacea, Dierama, Eupatorium, Crocosmia* and *Sedum* 'Herbstfreude'. Grasses will usually stand for a longer season than the perennials matched with them, so later in the season, you may have to cut the latter back to keep things tidy.

The options are almost endless, so take the following suggestions as a starting point and try out your own combinations.

The boldly coloured flowerheads of Macedonian scabious (*Knautia macedonica*) contrast beautifully with the blond, feathery grass of Mexican feather grass (*Stipa tenuissima*). Both *Heuchera* and bronze New Zealand hair sedge (*Carex comans* – a sedge, rather than a grass) are happy in shade, and work well in containers. The ruddy tones of the *Carex* complement the leaves of one of the bronze heucheras, such as *Heuchera villosa* 'Bronze Wave'.

A

Grasses are probably at their most striking used in bold groups, and often work well in difficult sites such as steep banks, where they can be used en masse. They can also mix with other plants to wonderful effect, provided that you make your matches carefully.

The aptly named hare's tail grass (*Lagurus ovatus*) goes well with verbenas. The fluffy, tufted heads look ideal against the erect stems and purple flowers of Argentinian vervain (*Verbena bonariensis*). The clean white-and-yellow stars of classic ox-eye daisies (*Leucanthemum vulgare*) look pretty mixed in with one of the smaller molinias, such as purple moor-grass (*Molinia caerulea* subsp. *caerulea* 'Moorhexe'). One of the aptly named fountain grasses, *Pennisetum orientale*, makes a striking pairing with any of the purple alliums.

The purple flowers of *Allium komarovianum* work well with oriental fountain grass, *Pennisetum orientale*.

Q How can I get instant oomph?

SOMETIMES YOU CRAVE SOMETHING that will make an immediate impact in your garden, whether you've arrived at a brand-new, bare-looking space or your garden already has structure or shape but isn't looking its best. Are there any ready-made solutions for speedy, good-looking results?

If you buy a lot of large plants that will soon come into flower and plant them out in your garden, it will cost a lot but it will deliver impact. Bedding plants – which may include annuals or half-hardy perennials (or, sometimes, bulbs) – are used to make speedy, seasonal, good looking displays in public places and you can apply the same rules in your own plot. There are two drawbacks: it's an expensive way to garden, and you lose out on the anticipation of seeing whether something you've planned over

Hardy garden chrysanthemums offer colour late in the season, often flowering all the way through to November.

A For something truly immediate, you might consider a bold, quickly planted up container. The added benefit is that containers can be replanted as necessary through the year to keep them looking their best. You can cheat, too, by dropping plants in pots into gaps in a border, supporting them with bricks if necessary. But, at a price, you can also buy semi-mature plants, and, with a bit more forward thinking, bulbs will certainly give a high-impact effect for the amount of effort they require.

WHAT'S WHAT: ANNUALS AND PERENNIALS

Most people know – or think they know – that annuals last just one year in the garden, although they may survive by self-seeding; that biennials are planted in one year and flower in the next; and that perennials (*perennis* is Latin for 'many years') will go on for longer, some for many years. But it's not as simple as that, because the term 'perennial' covers a huge number of plants, many with different qualities.

Hardy perennials are tough enough to survive a British winter and reappear the following year. They include herbaceous perennials such as hydrangeas and herbaceous peonies, which die back over winter, but grow fresh foliage in spring, and evergreen perennials such as hellebores, which keep their leaves through the cold season. So-called 'half-hardy' or 'tender' perennials are plants that are natives of warmer countries, such as geraniums. In their home surroundings, they will live for some years without special treatment, but in a cooler climate they won't survive winter without being given extra protection, and may not survive a tough winter at all unless they're taken under cover and protected from frost and damp.

The African lily, *Agapanthus*, a native of South Africa, has deciduous and evergreen forms and may need some protection in chillier climates.

a slightly longer term will work as well as you hoped it would.

The alternative is to take the experienced gardener's route and plan the season ahead, planting in autumn for the following spring. This gives hardy perennials the chance to establish themselves before flowering, allowing you to enjoy the full benefit of bulbs and plan for tender perennials or annuals to fill spaces in spring.

Q Do I need to lift bulbs?

TRADITIONALLY, in temperate climates, some bulbs and tubers, such as tulips, gladioli and dahlias, were lifted and stored over winter before being replanted in spring. This was intended to ensure that they survived to flower the following year, but is it really necessary? Will they reappear if they're just left where they are in the ground?

A

Whether or not a bulb will survive if it isn't lifted depends both on the winter and the bulb. Some tougher types will happily overwinter and reappear next year, while others – notably dahlia tubers – simply aren't tough enough to cope and are unlikely to reappear if left in the ground. Lifting bulbs is not always the answer, either.

Some bulbs, notably tulips, will rarely do well after the first year, even if they're lifted and stored before being replanted. Some gardeners leave them in the ground but feed them after flowering, in the hope that the additional nourishment will beef them up and encourage them to reappear the next year. Others treat them as annuals, take them out after flowering, and simply replace them with new stock the following year. Tougher types of bulbs will happily stay in the ground and reappear the following year, and there is a wide range of bulbs that will persist for many years.

Cyclamen, camassias, fritillaries, snowdrops, crocus, daffodils and narcissus are all examples of bulbs that, if planted quite deeply (usually at least three times the bulb's height) in a situation that suits them, will naturalise and spread.

Many will do well in grass – in which case, early-blooming varieties are best, so that their flowering season is over before it's time to mow the lawn – under single deciduous trees, or as part of a woodland garden.

Dahlia tubers won't usually cope with a cold winter in the soil. Lift and store them before replanting the following spring.

Are any flowers foolproof?

MOST ARTICLES written about plant seem to assume that readers will be naturally gifted gardeners, but if you have the opposite of green thumbs, are there any flowers that come with a guarantee to thrive?

If the idea of germinating seed feels daunting, buy seedlings instead. Nurseries sell small pots or trays of everything from wallflowers to sweetpeas, which just need planting and basic care to give results.

Annuals aren't expensive. Sweetpeas grown up a cane wigwam will flower for up to four months – the more you pick, the more they'll flower. The canary creeper (*Tropaeolum peregrinum*) is another colourful option. Sunflowers come in the gigantic bright yellow single-flower type or the subtler – and shorter – multi-flowered sort that are actually more useful as supporting players in the garden. Bright, daisy-like cosmos can be planted in containers or simply sown generously as a drift in a bed.

There are exceptionally easy-flowering plants available. And some annuals are also extremely quick growers. There is a number of options if you want fast, unfussy results.

Canary creeper, *Tropaeolum peregrinum*

Plants that produce

Many herbs are hardy, easy to grow and good for containers if you don't have much space. Only try basil if you have a really sunny spot; it needs heat to thrive. Less fussy are the different mints: try some of the more exotic options, such as chocolate mint or lemon mint, as well as the ordinary garden variety, but keep any mint plant in a container – they're famously invasive. Thymes, sage, parsley and dill are also all very easy plants to grow. Fennel, whether you grow the green or the bronze variety, is not only great with fish and in salads, but is also a beautiful feathery plant that can grow very tall in just a month or two. Other undemanding options with edible results include lettuce, radishes, chillies and runner beans. Peas are a good, easy-care crop to grow with children: the mangetout types grow quickly, and fresh peas are a treat that can be podded and eaten on site.

Q Which plants will suit my garden?

GETTING TO KNOW a new outdoor space is always exciting, whether it's a tiny patio or a fair-sized garden. You can find a suitable place in the garden for most plants, providing they suit the climate, but if you're experimenting with plants that are new to you, it's a good idea to get familiar with the basics, such as the soil condition and how much sun the garden gets, to ensure that you have the right spot.

The pH of the soil tells you whether it is acidic, sometimes also called ericaceous, or alkaline, sometimes also called loamy. DIY soil-testing kits are available online or from any garden shop or nursery. Acidity is measured in pH units, with the lowest numbers being the most acid, and the highest ones the most alkaline. A pH of 7 is neutral, neither acid nor alkaline. The majority of garden soil is mildly acid, with a pH of between 6 and 7, and this will suit a very wide range of plants. Soil preferences are usually given on plant labels, so if you're growing varieties you haven't tried before, check before you buy. A few plants like a specific pH.

A It's worth testing the soil pH of a new plot, and also getting to know which parts of the garden get the most light at which times of day. Testing the soil helps to give you the information you need to choose plants. Knowing how much light different parts of the garden get ensures that you won't put shade- or sun-lovers in the wrong place.

Those that strongly prefer a fairly acid soil – pH 6 and lower – include rhododendrons, camellias, azaleas and heathers.

Whatever kind of soil you have in your garden, the most important thing is to keep it healthy and fertile, with regular feeds of organic matter such as home-made compost or leaf mould.

Most heathers prefer acidic soils, but they are otherwise tough – most happily tolerating cold, open and windy sites.

WHICH WAY DOES YOUR GARDEN FACE?

The direction in which your garden faces – its 'aspect' – tells you how much sun it will get and at which times of day. This is important in deciding where to place plants that need more or less light and heat. Bear in mind that the sun's warmth in the morning isn't as strong as in the afternoon, so plants that like real heat are usually best placed where the sun will get to them later in the day. If you're not sure which way your garden faces, take a compass and stand facing towards the garden, with your back to the house wall. If south is directly ahead of you, you have a south-facing garden, if north is ahead of you, it's north-facing, and so on.

- **South-facing**: If your garden faces south, the area at the back of the house will have sun for most of the day, while the back boundary fence will be shaded all day. Looking outwards from the house, the right-hand side will get the morning sun, and the left-hand side, sun in the afternoon and evening.
- **North-facing**: If your garden faces north, the right-hand side will be in shade in the morning while the back and both sides of the house are shaded in the middle of the day. The right-hand side and the back of the house will have sun in the afternoon and evening.
- **West-facing**: If your garden faces west, it will be mostly in shade in the morning, with sun arriving on the right-hand side at noon, and travelling across the back of the house and most of the garden through the afternoon and evening.
- **East-facing**: If your garden faces east, it will get the most sun in the morning, with shade arriving on the right-hand side at noon. By mid-afternoon and through to sunset, most of the garden will be in shade.

South-facing

morning

midday

evening

North-facing

morning

midday

evening

West-facing

morning

midday

evening

East-facing

morning

midday

evening

Q Why has my lavender gone leggy?

LAVENDER IS DESERVEDLY POPULAR with both gardeners and pollinators; it's straightforward to grow, it smells great and looks good, and there is a huge number of different cultivars to choose from. But although it's generally easy-going, it does have some preferred situations and it needs a certain amount of care to do well.

A

Lavender that grows leggy and straggly usually hasn't had a haircut. It needs pruning every year to stay in good shape. It can become woody at the base, and in heavier soils this can happen in a comparatively short time. Even with care, lavender plants don't last forever.

Unlike some fussier Mediterranean plants, lavender can tolerate heavy, wet soils, although ideally it prefers poor, chalky, alkaline soil. You can help it in heavier soils by planting it in raised beds or on a mound; if you're growing a lavender hedge, plant it on a ridge to help with drainage. It also does well in pots and containers.

Lavender should be pruned annually, after it has flowered, in late summer or early autumn. This is straightforward: cut back the flower stalks and about 2.5cm of the year's growth. You need to cut it close, but don't take it right back to where the stems become woody – it's important to leave some green for the plant to grow from next year.

Out with the old, in with the new: taking cuttings

When it does grow woody – that is, when the lower part is all dry branches, and the foliage and flower stems are confined to the very top of the lavender – the plant needs to be replaced. Ideally you shouldn't situate the new plant in the same spot.

Lavender is easy to take cuttings from, so you can make new plants from

FIVE GOOD LAVENDER TYPES

Here are some of the best lavender types to try, from the classic English 'Munstead' to the white-flowered 'Edelweiss'.

English lavenders (right). *Lavandula angustifolia* 'Munstead' is a well-loved old variety, and a popular choice for lavender hedges. It grows to a height of 45cm and a spread of 60cm.

***Lavandula* x *chaytoriae* 'Sawyers'.** This is a large lavender with silvery leaves paired with lilac-blue flowers. It has a height of up to 70cm and a spread of 120cm.

French lavenders (below right). *Lavandula stoechas* subsp. *stoechas* Lilac Wings 'Prolil' is a beautiful example – each purple flower spike looks as though it is topped with little lilac wings. It is a compact plant with grey-green leaves, and it has a height of 40cm and a spread of 40cm.

Dwarf lavenders. *Lavandula angustifolia* Miss Muffet 'Scholmis' is great for containers, border edges or a low hedge. It has violet-blue flowers, a height of 30cm and a spread of 50cm.

***Lavandula* x *intermedia* 'Edelweiss'.** This is a large, bushy plant with masses of white flower spikes, a height of 75cm and a spread of 90cm.

old stock. Cut non-flowering shoots around 7.5cm long from the plant in early summer; strip off most of the leaves, leaving a few at the top; make a clean cut just below a leaf node on the stem; then plant in a small pot full of potting compost with a little extra grit for drainage. Place the pot out of direct sun, and keep it moist but not too wet. When the roots have formed, you'll notice new leaf growth – showing that the cutting has taken.

What killed my clematis?

WHEN A CLEMATIS that has grown healthily and flowered reliably for years suddenly collapses and dies without any warning, what happened? Could it have been saved?

Clematis plants do sometimes suffer a complete collapse, and often the plant dies. Clematis wilt (described more precisely as stem rot), a condition that is caused by a fungus, is often blamed, but there may be other factors at play.

Clematis wilt is caused by the fungus *Ascochyta clematidina.* The large-flowered types of clematis seem to be more susceptible than those with smaller flowers.

Both large- and small-flowered types may suddenly wilt, however, and in studies, the fungus has not always been detectable. Some people believe that the wilting condition may be prompted by a less-than-ideal environment. Clematis prefer deep, moist soil and fairly shady surroundings, so if they are climbing up buildings and walls they may suffer by being in the rain shadow of the building and in some cases the wall they're growing on may become too hot. Ways to avoid rain shadow include siting the plant a little way out from the wall, mulching it generously and using other plants around the base to help to shade the roots.

What to do if your clematis wilts

Cut all wilted stems back as far as the start of healthy stem, and burn the cuttings (don't put them on the compost heap). Disinfect the secateurs or pruners you used, too. New, healthy shoots may appear, although this can't be guaranteed – sometimes the plant will die.

If the worst happens and you need to replace your clematis, ideally choose a new location, rather than planting the new plant on the same site. If you have a smaller garden and changing the site isn't possible, change the soil instead. Dig out the area where the original plant stood, remove as much of the soil as you can (disposing of the soil, rather than relocating it in the garden), and dig in plenty of new organic matter. As well as watering it, mulch a new clematis to help it establish itself.

In plants afflicted by clematis wilt, the foliage collapses and shrivels very quickly, from the top of the plant downwards.

TYPES THAT ARE LESS LIKELY TO SUFFER WILT

All of the below deciduous types of clematis are good choices for avoiding wilt. Of the following, *Clematis viticella* is the only one that is seemingly unaffected by, rather than resistant to, the virus causing wilt.

Austrian clematis, *Clematis alpina.* A tough clematis that has single flowers (four petals), in bell shapes, or open, like stars.

Downy clematis, *Clematis macropetala.* This is quite similar to the *alpina* species, but the flowers come in doubles or semi-doubles (more than four or five petals).

Golden clematis, *Clematis tangutica.* This clematis has bell-shaped, hanging flowers in a vivid yellow, and petals that have a thick fleshy texture like lemon peel. It is tough and vigorous.

Mountain clematis, *Clematis montana.* This is an energetic climber with a potential stetch of 12m; flowers can be singles or doubles, and although they're not as big as the 'large-flowered' (and more wilt-prone) species, they tend to flower very abundantly.

Purple clematis, *Clematis viticella.* This species includes varieties with single or double flowers, which are open or in hanging bell shapes.

Austrian clematis, *Clematis alpina*

Golden clematis, *Clematis tangutica*

Purple clematis, *Clematis viticella*

Q How do I take cuttings?

IT IS WELL KNOWN that plants have an amazing ability to form new young ones from small, separated segments of themselves. How easy is it to take cuttings from favourite plants to bulk up the stock of the garden? How do you do it?

A cutting is a piece of a plant – it can be a section of stem, a stem tip, a root or even a leaf – which is encouraged to root and thus make a new plant.

A

Cuttings are relatively easy to take and, provided you have enough sheltered space to keep them until they establish roots and start to grow, you can create plenty of new young plants. Layering is another technique that it's worth learning, too (see box).

Cuttings can be taken from most types of plant. You need a clean, sharp blade, small pots with drainage holes and some light, easy-draining material to pot the cuttings in. This could be a mix of potting compost and sharp sand, perlite or vermiculite.

Softwood cuttings are usually taken in spring, from young shoot tips, from deciduous shrubs or either hardy or tender perennials. Semi-ripe cuttings are taken from midsummer to mid-autumn, once shoots have a hard base. These cuttings are usually taken from climbers, evergreen shrubs, herbs and ground-cover plants, among others (see pp44–45 for the basic technique).

Cuttings from tender perennials, such as fuchsias, can be taken in spring and planted directly into a mix of potting compost and vermiculite.

LAYERING

Layering isn't as familiar a technique to many gardeners as taking cuttings, which is a pity, as it's simple to do. While not fast, it will often work on plants for which cuttings don't tend to succeed, including jasmines, magnolias and camellias.

There are variations, but in its simplest form, the gardener encourages a shoot of a plant to make its own roots without separating it from the parent plant. If you're layering a deciduous plant, you can do it in winter or spring; for evergreens, winter is best. You will need rooting powder, a short piece of thick wire, and a bamboo cane. The plant needs to be large enough to have a flexible shoot around 40cm long to work with.

▼ Layering works slowly but it produces a sturdy new plant, which should be well established within a year.

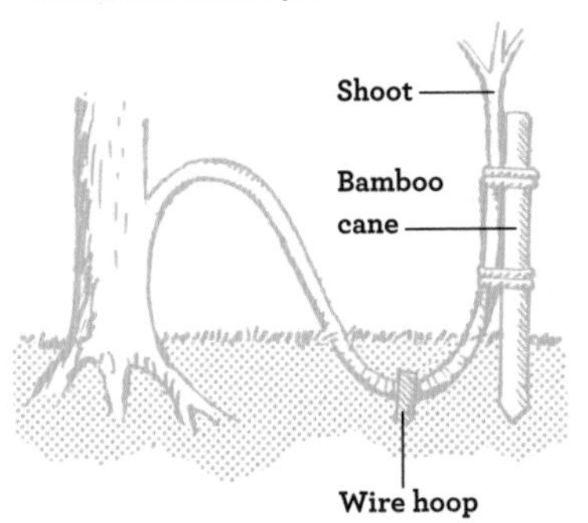

1 Pull a flexible shoot from the outside of the plant down to ground level and mark the point where it touches the ground with the bamboo cane.

2 Use a sharp knife to make a 5cm incision along the length of the stem. It should go through a leaf bud. Paint the surfaces of the cut with rooting powder.

3 Make a shallow trench about 10cm deep halfway between the bamboo cane and the parent plant, then peg the incised section of stem into the trench with the wire, bent into a short hoop.

4 Leave the tip of the shoot uncovered and attach it, pointing upwards, to the cane.

5 Refill the trench with soil over the pegged stem, firm the soil lightly and water if the ground is dry.

The layered shoot should have put down roots and turned into its own plant within a year, after which it can be separated from its parent and planted in a new position.

Q How do I deal with potentially harmful plants?

OCCASIONALLY PLANTS CAN DO US HARM – for example, the milky sap of euphorbias can cause a rash on some sensitive skins. But are any common garden plants really dangerous and, if so, do they need to be avoided completely?

In terms of accidental harm, your home is a far more dangerous place than your garden. Good gardening practice – wearing gloves, washing your hands after gardening and avoiding eating or drinking while handling plants or green waste – will protect you in the course of everyday contact with most plants, even toxic ones.

A Very few plants in the average garden can do you serious harm unless you eat them. It's a good idea to be respectful of different plants' various qualities, and children and animals need to be protected from toxic plants.

Know your enemy

Most toxic garden plants are only really dangerous if they are eaten. Small children can't be left unsupervised in the garden and should be taught as soon as they are old enough to understand that they must not put random leaves or flowers – or any other plant material – in their mouths. Pets can't be taught this, but most dog or cat owners know if their pet likes to eat plants (many never show any interest) and should keep an eye out accordingly.

Plants that are highly toxic when eaten include yews (*Taxus*), foxgloves (*Digitalis*), shrub verbenas (*Lantana*), monkshood (*Aconitum napellus*) and the castor oil plant (*Ricinus communis*). You need to know what the plants in your garden are, and to make sure that children or pets are kept away from any potential

Monkshood, *Aconitum napellus*, is highly toxic – but only if eaten. Wearing gloves will protect you from any ill effects.

problem – if there's anything that really worries you, get rid of the plant. And be particularly cautious around plants with bright, toxic berries, such as yew or cuckoo pint (also known as lords-and-ladies) – these can have an obvious initial appeal to small children because of their resemblance to sweets, although most don't taste good to humans.

Apart from those plants that are poisonous when eaten, other potential problems include those with tiny hairs that can irritate the skin or, in a few cases, cause a bad burn if combined with exposure to sun.

If there's a problem

If you have reason to think a child has eaten any part of a toxic plant, take him or her to hospital, along with a piece of the plant you think they've eaten. Don't try to make the child sick. In the case of a pet, do the same – but head to the vet rather than the hospital.

WHY CAN BIRDS EAT YEW BERRIES WITHOUT BEING POISONED?

Birds love berries, and it is common to see a blackbird or thrush dining off the red fruit of the yew tree without seeming to suffer any ill effects. This is because the birds don't actually break down the seed, which is the only poisonous part of the berry. They eat the sweet berry flesh and the seed passes straight through, leaving the bird unharmed.

English yew, *Taxus baccata*

There are other plants, though, that wild birds seem to be able to eat and humans can't. One is hemlock (*Conium maculatum*), which contains coniine, a poison that has a disastrous and often fatal effect on the human nervous system. Many wild birds, however, seem to be able to eat hemlock with impunity – quite why is not completely understood. Although the coniine has no effect on them, it does build up in their systems and cases of coniine poisoning in humans have been reported in Italy, where songbirds are hunted and eaten, as recently as 2007. Doctors concluded that it was the coniine stored in the birds that, in turn, had poisoned the humans who dined on them.

Q How can I avoid garden thugs?

IT'S IRRITATING ENOUGH when you're plagued by persistent weed pests like ground elder and horsetail, but it's even more annoying if you introduced the unwelcome incomer yourself. Beware of strangers, at least if they're of the plant variety – make sure you've been properly introduced before you let them put down roots.

Most garden thugs are pests by virtue of their sheer, unstoppable vitality – plants that grow too fast, self-seed or perpetually and rapidly send out underground roots, suckers or bulbs. Picking out weed seedlings can become a constant chore. For example, horseradishes grow incredibly deep roots, and a Russian vine planted to cover an ugly fence may look pretty, but it will soon prove that one of its common names, mile-a-minute, is an almost literal description of its growing speed.

A

Do your homework and check before you buy a plant, taking descriptions from several sources and, if possible, talking to someone who already grows it. An innocent-sounding 'excellent ground cover with good spreading habit' can become 'invasive nuisance' quite easily when you get to know it a bit better.

Exerting control

Frequent heavy pruning can sometimes keep things in order above ground; below ground, doing battle with invaders like bamboo can be a long fight. Once a garden thug is in place, you have to do the best you can to eradicate it.

A pervasive plant can be one of the few arguments for the use of chemical control, with systemic, permitted herbicides. If you do decide to go the chemical route, research properly before use, and use only as much as is absolutely necessary to rid you of the thug. Non-chemical fixes may include stifling, using black plastic sheeting heavily weighted down

Pampas grass, *Cortaderia selloana*, is easily controlled in the UK, but is considered an invasive pest in many states in the USA.

Japanese knotweed,
Fallopia japonica

BETTER SAFE THAN SORRY

Sometimes it's best to guard against breakout from the start. Some smaller runaway plants, such as mint, can be controlled in pots, used either as conventional containers or sunk into the bed or border to form a root barrier that you can't see. In order to grow appealing but invasive plants, such as many members of the elegant bamboo family, some gardeners go to the lengths of confining them in large pits, lined with black root-guard plastic.

Corn mint,
Mentha arvensis

over the soil surface – and, in extreme cases, with a trench dug around the area and also lined with plastic, to stop below-surface breakouts – or meticulous, but inevitably rather time-consuming, digging out.

Disposing of the body

Don't make the mistake of putting green waste from a garden thug on the compost heap. If there isn't too much of it, it can be burned on a bonfire; if there is a large volume of waste material, or if you don't have a suitable bonfire site, you can take it to a green waste centre for composting. In a some instances there is specific legislation dealing with the way in which the green waste from notorious plant pests may be disposed of – for example, any waste from Japanese knotweed is not allowed to leave a domestic garden unless it goes straight for disposal at a licensed landfill site.

Q Which flowers last longest?

ARE THERE ANY PLANTS with exceptionally long flowering seasons? In fact, are there any that will go on flowering for most of the year?

A The perennial wallflower is exceptionally long-flowering – but it is pretty much one of a kind. Most plants have adapted to spend the amount of time in flower that will get them the best results in terms of pollination. Try double-flowered plants, which can't be accessed for pollination as easily as other types and tend to flower for longer.

Flowering and setting seed are an exhausting business for a plant, draining its resources and leaving it without much to draw on. Given this, it's important that it gets it right – spending most of the year in flower would be too exhausting to maintain and wouldn't leave a plant with enough energy to reproduce itself. Generally, once fertilisation has successfully taken place, it stops flowering and gets on with making viable seed to ensure future generation.

The longer-lived exceptions to the rule are plants with 'double flowers', which have many more petals than single-flowered types. While they may still attract pollinators, often they don't have a structure that allows for pollination, encouraging them to flower for longer. Some new variants have also been bred to be sterile, meaning that they cannot reproduce. If gardeners would like more of these appealing but sterile plants, they will have to buy them or take cuttings, or divide the plants they already have.

An appealing rust-coloured variety of yarrow, *Achillea*, whose flowerheads will last all summer, only dying away in late September.

EIGHT PERENNIALS THAT FLOWER FOR AGES

Coneflowers, *Rudbeckia*. Bright yellow daisies with dark centres, in flower from June until November.

Penstemons, *Penstemon*. Spires of small, foxglove-like flowers in a range of colours including pinks, reds, purples, blues and creams, in flower from June until November.

Macedonian scabious, *Knautia macedonica*. A dark-red scabious with pincushion flowers. It flowers from early July to the end of September. Most scabious are long-flowering, so consider some of the white or blue versions, as well.

Spurges, *Euphorbia*. Many euphorbias flower early in the year, before there is much other colour about, and have a three-month-plus flowering season. There are types to suit almost any situation.

Greater masterwort, *Astrantia major* 'Alba'. The white form of the pretty, spiky *Astrantia*, which also comes in pinks, reds and mauves. Astrantias are happy in shade, and flower from May to October.

Purple coneflower, *Echinacea purpurea*. The most common of the coneflowers, with a prominent, rust-coloured central cone and a purple, daisy-like flower. Will flower from July to October, provided it is deadheaded regularly.

Yarrows, *Achillea*. A large group of unfussy perennials, with feathery foliage and flat flowerheads that come in a wide range of colours from white and pinks through shades of yellow, terracotta and red. Will flower from June through to September.

Geraniums, *Geranium*. Some perennial geraniums have extra-long flowering periods. Three good ones to consider are Rozanne 'Gerwat' (pictured right), 'Anne Folkard' (vivid magenta) and 'Orion' (mauve-blue).

Q Why won't my flowers grow on the right side?

WHY HAS MY ROSE, which flowered happily in its first year, migrated to the other side of the garden fence? Passers by are now getting the benefit of its flowers, and there are none on the garden side.

Japanese snowball 'Rosace'
***Viburnum plicatum* f. *plicatum* 'Rosace'**

A

It's not that the rose is flowering on the wrong side, it's that it's been planted in the wrong place. Climbers are opportunists who will grasp any chance to maximise the available light, and your rose is simply doing what comes naturally.

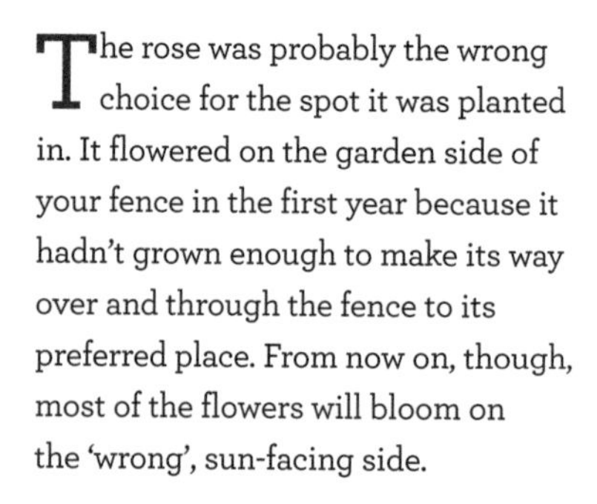

The rose was probably the wrong choice for the spot it was planted in. It flowered on the garden side of your fence in the first year because it hadn't grown enough to make its way over and through the fence to its preferred place. From now on, though, most of the flowers will bloom on the 'wrong', sun-facing side.

Match the plant to the site

If the spot is too shady for a rose, choose an alternative that will be happy where it is. Consider camellias, mahonias or viburnums instead – and evergreens may be the best choice if the site is really shady – they will still look good in winter. Anywhere with an uneven pattern of light will tend to result in uneven growth in the plant.

Common camellia,
Camellia japonica

Q What if I don't like flowers?

FLOWERS seem to be the constant focus in popular gardening programmes and magazines – and whenever anyone talks about gardening. But what if your preference is for greenery – can you have a garden without concentrating on flowers?

Taking flowers out of the equation can lead to some interesting styles of gardening. The recent popularity of Japanese-style gardens has broadened ideas about what constitutes a 'green' garden – look to acers, maples, ferns, grasses and even mosses for appealing results that aren't dependent on obvious blooms.

Hay-fever sufferers may end up disappointed if they believe that eliminating flowers in the garden will alleviate their symptoms, though. Most irritations are caused by the pollen of grass and trees, and pollen is a distance traveller – the source of the irritation may be travelling from much further away than your garden. It may be worth researching pollen-rich plants that are known irritants and avoiding them, but don't expect a magic cure.

Looking at structure

If you take the focus off flowers, you may be surprised by how much colour and variety foliage can supply. Try judging plants by their leaves and overall shape instead of their flowers. Ask questions about foliage as well as flowers when you buy plants – how well do the leaves last, and will the colour change through the season? What will they look like when they die back? The answers will tell you what part the plant might have to play in your garden.

If you have the space to consider trees, consider acers, eucalyptuses or rowans, for their standout foliage, or silver birch, for bark.

In beds and borders, consider mixing the fine leaves of grasses with the broader leaves of other perennial plants. Shrubs such as *Artemisia* and *Pittosporum*, too, have foliage that greatly outshines their flowers.

For plants to use as ground cover, a large number of pulmonarias, hostas and epimediums have pretty leaves that will do the job very effectively.

A

You can easily avoid having flowers. Despite the popularity of flowering plants, most gardens rely on a green background and structure. Thinking about this part is a useful exercise for anyone interested in garden design.

Chapter 2

Edible Plants

Q Why are my carrots curly?

Sometimes home-grown vegetables can take a very individual form. Cross-legged carrots, split potatoes and oddball tomatoes may all feature in the vegetable patch. What causes curly carrots – and how do you know when they're fine to eat, and when you should leave well alone?

A Vegetables often end up oddly shaped because they have been grown in lumpy soil – but they will still taste good. Many supermarkets even sell a range of oddball vegetables now. Sometimes, though, the misshaping can be caused by pests and diseases, meaning the crop may not be fine to eat.

Twisted carrots may take slightly longer to prepare in the kitchen, but will taste perfectly good.

Carrots can stunt or deform if they, as tiny seedlings, have had to force their way through poor or lumpy soil, or if they are first crowded and then thinned roughly. It's a myth, which you still sometimes hear, that heavy manuring leads to forked carrots. Although both carrots and parsnips are often sold as plug plants in module trays ready to be planted directly in the vegetable patch, experienced gardeners will walk straight by these offerings. They know that transplanting the seedlings like this almost always leads to disastrously distorted roots.

Other problems

Splitting is a different problem, causing vegetables, such as potatoes, beets, radishes and carrots, to be split down their length, revealing the core. Potatoes may also form 'dollies' – odd-shaped protuberances. Although still edible, preparing split vegetables is more time-consuming, and a lot of the root may go to waste. The cause

CAT-FACED TOMATOES

Misshapen root crops aren't the only malformations in the vegetable garden. So-called 'cat-facing', the condition that causes puckered indentations and abnormal growth in tomatoes, is another common problem. This has a variety of causes, including too-low temperatures while the crop is developing and too-high nitrogen levels in the soil. The fruit is edible, though, if the damage is cut away.

is often an erratic amount of water available to the vegetables as they grow; if the soil around them is allowed to dry out before subsequent heavy rain or saturation watering, they'll 'drink' more than their structure can take, and split as a result.

Carrot damage that is caused by nematodes – which can be diagnosed when the root growth turns yellow – can cause not only stunted roots but also woody centres. Aphids, too, can bring viruses to a healthy crop. Probably the most damaging of all the pests is carrot fly, whose larvae ruin carrots and other root crops, making numerous tunnels through them and rendering them inedible. The fly is attracted by the smell of young carrot foliage – which is stronger when the leaves are crushed – so avoid rough or careless weeding, which may entice them to your crop.

Splitting can cause the exposed part of the carrot to go dry and 'woody', and may mean much of it goes to waste.

Getting a good crop

Give your carrots the best chance of growing into model specimens by preparing the bed meticulously and giving them plenty of space. You can opt to plant early, ideally avoiding the drought and stop-go water supply that can be problematic later in the year. If your soil isn't ideal for carrots – that is, if it is heavy clay, stony or shallow – choose stump-rooted varieties. As the name implies, they have naturally blunter ends which are less liable to deform or distort than fine-rooted carrots as they grow through less-than-ideal soil.

Protect the developing plants with fine mesh or fleece to avoid both virus-carrying aphids and carrot fly.

Nematode problems are best avoided by practicing crop rotation, and ideally only repeating the same crop in the same space at most every third year.

Q How can I make my vegetable patch look prettier?

THE CONVENTIONAL VEGETABLE PATCH, with its neat straight lines, can be a thing of beauty in its own right. But what if you want something rather less strictly arranged – how can you indulge your flair and creativity in the vegetable garden?

The original reason vegetable gardens were so tightly organised was to save on work. Regimenting crops in neat, straight rows and tidy, self-contained beds makes them easier to weed. Leaving enough space to walk between the crops ensures that you will not tread and compact the soil of the beds themselves. Most of today's gardeners cultivate relatively small vegetable plots, so there is no reason you can't plan yours as creatively as you do the rest of your garden.

Visual planting

When you're choosing the season's crops, consider how they'll look as well as how they'll taste. This can include all kinds of aspects of the plant you're choosing – the colour of its leaves and flowers, its spread and its height.

Choose appealing supports for climbing crops – beans, peas and other climbers will be just as happy to scramble up them as they will ordinary cane-and-string arrangements. Arches, willow wigwams and wooden obelisks will all look good and create focal points among lower-growing crops.

A Beauty is in the eye of the beholder, of course, and plenty of gardeners rejoice in a well-regimented and productive vegetable patch. Otherwise, you can consider tall plants for structure, crops with bright flowers and pretty supports for your climbers.

Large, structural-looking plants, such as the thistle-like globe artichoke, will add height for visual impact as well as being good eating.

Globe artichoke,
Cynara cardunculus
(Scolymus Group)

COMPANION PLANTING IN THE VEGETABLE GARDEN

Companion planting is the process of siting one plant near another because it benefits it in some way. The best-known example is that of the 'three sisters', the classic Native American combination of beans, maize and squash. Strictly speaking, this particular trio are intercropping; each plant has a different growing habit, making full use of areas both above and below the ground – the beans use the maize as a support, while themselves enriching the soil with nitrogen, and at ground level the squash has foliage that shades the soil, guarding its moisture and discouraging weeds.

▲ The 'three sisters', beans, maize and squash, form a mutually beneficial partnership in the vegetable plot.

Certain other plant mixes are widely believed to work well together. The claims aren't necessarily supported by science, but see if they work for you. Plant companion crops with discretion, though – you don't want them to thrive to the point at which they'll compete with your main, edible, crop. At the least, you'll be creating attractive vegetable and flower or herb mixes with the following plant combinations in your vegetable patch:

- **Garlic chives and carrots** – enthusiasts believe that the chives confuse carrot fly and keeps it away from the root crop.
- **Lavender and marigolds** are both said to deter greenfly, so plant either or both alongside any crop that has proved particularly aphid-prone. (Some gardeners also believe that nasturtiums will attract aphids away from beans.)
- **Borage**, with its bright-blue flowers, is said to be particularly appealing to pollinators – good news for any vegetable plot.

Q Is home-grown tastier?

ASSUMING that your vegetable garden isn't large enough to grow every single variety that might appeal to you, which vegetables are the ones most worth growing? Are there standouts, as far as home-grown taste versus store- or market-bought flavour is concerned?

The produce that you can buy in the shops will inevitably have had to travel some days through the supply chain, which won't have helped its freshness or flavour, giving home-grown crops an advantage. Some home-grown crops, however, such as apples, pears, potatoes and carrots, may show a less marked superiority.

Garden pea,
Pisum sativum

Size matters

Growing your own offers you the opportunity to harvest the crop at its best and eat it immediately. Some commercial crops are never allowed to grow to their full size – supermarket parsnips, for example, are always harvested when comparatively small. If you grow your own, you can let them get larger: they may look more gnarled and less Instagram-worthy, but they will be just as tasty as their miniature siblings. The types of potatoes grown commercially also tend to be selected for uniformity of appearance and reliable cropping, so if you want to try some of the more unusual tastes, textures and colours that are very widely available as seed potatoes, it's well worth growing your own.

Little and often

Planning your planting allows you to gather and eat your crops at their peak. Salad, peas and beans are all best harvested and eaten little and often, so many gardeners opt to stagger their sowing and planting to get as long a season from individual crops as possible.

Vegetables in which sugar quickly converts to starch after harvesting will taste best if you can eat them almost immediately – true sweetcorn enthusiasts recommend that you have water already on the boil before you cut the cobs from their stems. Bought cobs will unavoidably be less fresh, and won't taste so good. In the

A

To some extent, what you prefer depends on personal taste. As a general rule, though, produce with a short shelf-life – salads, fresh beans and peas, and soft fruit – is tastier direct from the garden.

commercial world, peas grown for freezing are mechanically harvested and frozen within two hours. This means that, in the store, frozen peas are a better product than fresh – but not in the garden, provided that you can get them from plot to pan in a couple of hours or less. Or, of course, you could eat them directly from the plant, when they're at their most delicious.

Tastier tomatoes

It is common knowledge that home-grown and heritage tomatoes tend to taste better. The University of California found that over the last 70 years, the commercial breeding of tomatoes to produce fruit that ripens evenly, and is thus easier to transport, has inadvertently suppressed the gene in the fruits that contributes to sweetness. Breeders are countering this by producing 'neo-heirloom' cultivars to their full size, with the aim of maintaining flavour but allying it with both high yields and pest and disease resistance.

COURGETTES: BOTH FRUIT AND FLOWER

Some benefits of home-grown aren't available commercially. For example, generally in colder climates, it's unusual to see courgette flowers offered for sale alongside courgettes, although they're a common sight in southern Italian or French markets. The flowers are a delicious bonus for the home courgette grower.

Incidentally, if you do pick your own courgette flowers, unless you have such a heavy crop that the flowers can be spared regardless, make sure you choose the male flowers which won't go on to produce fruit. Male flowers have a narrow stem and also tend to open earlier in the day. Female flowers have a swollen stem that, after pollination, will develop into a courgette.

A male (left) and female (right) courgette flower. These are delicious stuffed, in a salad or battered and fried.

Q How can I grow super-sized vegetables?

IF YOU'VE EVER visited a local flower-and-produce show and been amazed at the displays of truly vast vegetables, you've probably wondered how the results are achieved. You may even have harboured some secret ambitions in this direction yourself.

While the majority of classes at local shows reward evenness and quality of vegetable exhibits over pure size – and, to be honest, these are a better test of growing skills than sheer heft – an attempt to grow a record-breaking, or at least a show-winning, pumpkin can be an entertaining project.

A

Super-sized is often a hobby of long-time gardeners. Huge pumpkins and marrows are the most commonly seen, as, unlike with giant leeks or onions, they don't require a greenhouse. You will need the right seeds and soil, some space and patience.

THE HYDE HALL RECORD-BREAKER

In 2016, a single pumpkin seed fetched £1,250 at auction. It came with an impressive pedigree – from Beni Meir, a Swiss grower, who had raised record-sized pumpkins in greenhouses. This seed was to be grown outdoors instead: raised tenderly by Matthew Oliver in the RHS garden at Hyde Hall in Essex, it did not disappoint. The resulting pumpkin weighed in at a stunning 605kg.

RAISING AN OUTSIZE PUMPKIN

- Choose the right seed. Some varieties grow larger than others, and although there's a clandestine trade in the seeds of known performers between experienced gardeners, some suppliers also offer high-quality seeds with the potential to grow super-sized fruit.
- Start by sowing the seed in on its side, in a seed mix, in April. Keep the pot warm – at least 18°C – ideally in a propagator.
- Harden the seedlings off, either in a cold frame or outdoors, in May. Move them indoors overnight, then plant them out two weeks later in early June, ideally under a cloche or fleece.
- Pumpkins need both space and fertile, free-draining soil – allow 1.8m between plants, plus a bucket of well-rotted manure per square metre and add regular fertiliser feeds. If you don't have much room, try planting a single pumpkin in a raised bed, or on a mound of well-fertilised soil around 25cm high. If you have space for more than one, grow several and pick out your potential champion when they've started to become established.
- Make a planting pocket for each seedling about 30cm deep, and fill it with a mix of garden compost and soil.
- Keep the soil moist but never soggy, to allow your pumpkin to put on size and to avoid mildew. Sink a 15cm pot alongside each seedling and water into this. This stops the soil surface getting too wet, and ensures the water goes down to the roots.
- If you hand-pollinate the plant the fruit should 'set' early. Use a paintbrush or your finger to transfer pollen from the pointy stamen of the male flower (the male has a narrower stem) to the central, sticky-looking stigma of the female flower (this will have a wider, swollen-looking stem).
- Once the fruit has started to develop, feed the plant every two weeks with a high-potassium potash fertiliser.
- Place each pumpkin on a piece of tile to keep it clear of the soil; this will help to avoid staining and perhaps rot. Turn the fruit from time to time to encourage it to swell evenly – at least until it gets too big to handle.

Q Why have my tomatoes got black bottoms?

YOUR TOMATOES have developed dark hollows and spots at the end of the fruit furthest away from the stalk (at the point where the flower fell away from the fruit). What's wrong, and can it be prevented?

A This is a condition called blossom end rot, and it happens because the fruits are short of calcium. Although it most commonly affects tomatoes, it can occasionally strike with peppers and aubergines, too.

The degree to which the problem affects the fruit varies from a smallish dark spot to a large, softened rotten-looking area that may take up most of the base of the fruit. And while you can cut away the spoiled part and eat the rest, in the worst cases there may not be much 'rest' to eat.

Although it is caused by a lack of calcium, this doesn't usually mean that the soil is low in calcium, but that the plant can't move calcium fast enough to its extremities – where the fruits are – to keep up with its rate of growth. The main cause is irregular watering. Plants need even, regular watering to ensure that calcium is constantly being carried through the plant in the water, then lost by evaporation from the leaves and fruits.

When crops are being grown under glass, it's also important to keep them well ventilated, as very high humidity levels can result in limited evaporation, and thus also limit the flow of water through the plant.

At the first sign of blossom end rot, ensure that your watering regime is regular and – if in the greenhouse – try to increase ventilation.

Can soft fruit bushes live longer?

Do soft fruit bushes always have a limited productive life, or can they be coaxed to continue to crop well into old age?

Soft fruit is rewarding to grow, as it is fairly easy to get a good crop and most varieties of soft fruit are relatively expensive to buy in the shops. How long the plants last depends on which fruit you're talking about – raspberry canes, for example, or gooseberry bushes may have a long life: some gardeners find that they're still productive after well over a decade. Strawberry plants (*Fragaria × ananassa*, right), in contrast, will need replacing every two or three years.

Fresh stock

Whichever type of soft fruit you're talking about, the bushes or canes tend to fall victim to viruses and other diseases and pests after a number of years, and cropping will usually then fall off drastically. And because new cultivars are constantly being introduced, many excellent performers and certified virus-free, you should harden your heart and replace older bushes that are no longer top performers.

Just as with vegetable crops, it's a good idea to practice rotation with soft fruit, so don't put new plants in the same location as the old ones. Even if you have to relocate a fruit cage, it will probably be worth it – planting new stock in a fresh spot usually results in startling increases in yield. And the ground where you've grubbed up the old fruit bushes will have soil that is already in excellent condition for planting vegetables.

New or old?

Should you go for new cultivars, on the assumption that they're likely to stay disease-free and will probably also crop better? While it's true that modern cultivars are usually better than the old ones, traditional varieties may still give good value. The traditional 'Baldwin' blackcurrant, for example, remains hard to beat. If your space is limited, read up on varieties carefully and pick the one you like the sound of most.

Soft fruit bushes and canes can live for decades: allotments are full of aged examples that have passed through many owners. Whether or not they will still crop well or reliably, though, is a different matter.

Q Does my vegetable garden need paths?

ARE PATHS CRUCIAL in a vegetable garden? And if they are, what are the best ways to make them, and which materials should I use?

A Traditionally, vegetable gardens didn't have paths – gardeners simply walked up and down on the soil between the rows. But it's always worth trying to reduce the amount of trampling, and paths are necessary if you want to practice the popular no-dig system.

Permanent narrow beds (between 1 and 1.5m wide) can be worked entirely from the paths between them – you shouldn't need to walk on the soil of the beds at all. If the soil isn't trodden down, the maintenance work will be reduced, and it won't always be necessary to dig the beds. However, if you live in an especially wet area, or if your soil is slow-draining clay, raised beds may be the best option.

What kind of path?

If you want – or need – paths in your vegetable plot, what's the best material to make them from? It all depends on how much time and effort you're able to put into making them, and whether or not your veg garden is going to be kept on the same site long-term.

The easy options are bark or chipped wood or – less satisfactorily, because it lasts only a short time – straw. These materials are quick to put down and when they rot can be added to the soil, but they need frequent replacement.

Brick or bark make the most effective paths. Bark is easy but will need replacing quickly; brick lasts, but costs much more.

WHAT IS 'NO-DIG'?

Ever since people began growing vegetables, they've looked at ways to avoid the irksome task of turning the soil. No-dig is a system, developed during the 1980s and 1990s, for maintaining a vegetable garden that doesn't depend on the traditional heavy digging. With beds kept narrow enough to tend from paths alongside them, no-diggers use a thick layer (between 5 and 8cm) of rotted manure or compost laid on top of the soil to smother weeds, but don't dig it in. Worms relish the mulch and their burrowing and soil processing enhances the soil structure.

No-dig is good for people who, for whatever reason, can't manage heavy digging, and also when the soil is a heavy clay that is only workable for brief periods in spring and autumn.

At first, many gardeners found it hard to believe that crops could flourish if the ground wasn't thoroughly dug over, but gradually the system became popular. When done well, it is very effective, although the full benefits may take several years to show. It's worth persisting: enthusiasts praise it as not only requiring less labour but also more environmentally friendly. Charles Dowding, whose writings have done a lot to make it currently popular, calls it 'working with, rather than against, the soil'.

The hard-work choice is brick or pavers (new or reclaimed) – either is relatively expensive and will take much longer to lay, but both are easy to work from and to clean, and may well be worth the extra work at the start if the garden is a permanent project.

Gravel should be avoided – it gets claggy when mixed with soil and mulch and can't be effectively cleaned.

Laying an attractive brick path doesn't need to be a job for the professionals: stick to a simple pattern and tackle the project yourself.

Q Is crop rotation necessary?

CROP ROTATION has been around for centuries – at least since Roman times, if not longer. It is based on the principle that crops with similar nutrient needs – and susceptibilities to similar problems – should be moved around, rather than grown in the same ground each year.

In rotation, crops are grouped by family, and each group is grown in a different part of the vegetable garden each year. Plants of the same group are moved through the garden plot year by year, only returning to a previous site once every five years in a five-year rotation cycle – or four years in a four-year rotation. Rotation minimises the chance of plants suffering from crop-specific diseases, and also of pests becoming too entrenched in areas where particular crops are grown repeatedly. The soil is also less likely to become depleted of the specific nutrients favoured by some crops.

Preventing problems

Three of the most serious issues that crop rotation combats are the potato cyst nematode (in potatoes), club root disease (in brassicas) and white rot disease (in the onion family). These problems can all survive in the soil at damaging levels for years, so it's worth trying to prevent them from appearing. A three- or four-year rotation may only provide partial protection, but it is still well worth having.

Bear in mind that some crops don't suffer much from root diseases or pests, so can be used as fillers anywhere you have space – they include pumpkins and squash, sweetcorn, courgettes, sweet potatoes, lettuce and endives.

If your garden really is too small to use a rotation system, follow these three simple guidelines:

1. Keep your crops well mulched.
2. If any crop has a bad attack of a pest specific to that crop, avoid growing it again in the same place, or at all if your garden is really tiny, for a year or two.
3. Keep your vegetable garden clean, and clear away all dead foliage at the end of the growing season.

A There are excellent scientific reasons for crop rotation, but home gardeners seldom have enough land for the very long – five years or more – rotations that give the very best results. Even on a small plot, though, a three- or four-year rotation is usually possible. Most home gardeners find a rotation system helpful in raising healthy crops.

THREE- AND FOUR-YEAR ROTATION PLANS

Start by putting the crops you want to grow into groups, as follows:

- **Brassicas** – including cabbage, cauliflower, kale, kohlrabi, Brussels sprouts and, perhaps surprisingly, turnips and swedes.
- **Potato family** – including potatoes and tomatoes; although bell peppers and aubergines also belong to the family, they suffer from fewer problems and so can be slotted into any gaps in the plot as there is space.
- **Legumes** – including peas and broad beans; as above, although French beans and runner beans belong in this group, they're generally relatively problem-free, so can be slotted into the plot anywhere they fit.
- **Onion family** – including onions, leeks, shallots and garlic.
- **Root vetables** – including carrot, beetroot, celeriac, celery, Florence fennel, parsley and most other root crops apart from turnips and swedes.

Here's how to plan both a three- and a four-year rotation:

Area A
Year 1: Potatoes
Year 2: Legumes, onions, roots
Year 3: Brassicas

Area B
Year 1: Legumes, onions, roots
Year 2: Brassicas
Year 3: Potatoes

Area C
Year 1: Brassicas
Year 2: Potatoes
Year 3: Legumes, onions, roots

▲ Three-year rotation in three areas, A, B and C, with potatoes and brassicas as the leading crops.

Area A
Year 1: Legumes
Year 2: Brassicas
Year 3: Potatoes
Year 4: Onions, roots

Area B
Year 1: Brassicas
Year 2: Potatoes
Year 3: Onions, roots
Year 4: Legumes

Area C
Year 1: Potatoes
Year 2: Onions, roots
Year 3: Legumes
Year 4: Brassicas

Area D
Year 1: Onions, roots
Year 2: Legumes
Year 3: Brassicas
Year 4: Potatoes

◄ Four-year rotation in four areas, A, B, C and D, with onions, roots and legumes playing a more important role.

Who's stealing my apples?

If you've invested time and care in your apple trees and managed to keep the fruit free from disease, it's frustrating to lose some of that treasured crop to an apple thief. Take a careful look at the evidence to identify the most likely culprit.

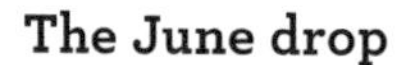

The June drop

First, what time of year is it? Trees use their natural good sense to jettison apples if they have a crop that's too heavy for them to carry to ripening. Although this is known as 'the June drop', it usually takes place in July: you'll find the ground under the tree littered with unripe fruit. Apple sawfly, a pest whose larvae tunnel through

BAGGING APPLES

One way to keep pests and diseases away from your ripening apples is to bag them on the tree. This isn't a new technique, but it is becoming increasingly popular with hobby fruit growers – you thin a young apple cluster to just one fruit, then open the centre of the zip-top of a plastic food bag, put the apple in and fasten the bag around the branch with a twist tie. Snip one of the lower corners of the bag to allow moisture to escape, and leave the fruit to ripen. Users report large, blemish-free fruit, so it's worth trying, although it is quite labour-intensive to bag the whole tree.

In the past gardeners would sometimes grow an apple or a pear in a bottle. Simply place a bottle over a fruit on the branch while the fruit is still small, then, when the fruit is ripe, detach it. It will appear to have passed – impossibly – through the bottle's narrow neck. There's also an alternative idea that will give you a fruit that's shaped like something else. You can buy plastic moulds that are fastened over the growing fruit to make a perfect cube shape, say, or even a miniature Buddha, offering an amusing novelty for children and 'how-did-you-that?' appeal for adults.

the young apple, can also cause an early-season fall of fruit (between late April and June).

However, apple trees usually set a heavy crop, so these early losses can generally be ignored as the crop matures. And it's at this point that other apple thieves may make themselves known. Look at the clues to see who's to blame.

Apple thieves tend to be seasonal, but can include squirrels, birds, deer, badgers and even humans. And sometimes a thief isn't involved at all – natural wastage may play a part.

Who stole the apples?

Nibbled fruit on the grass? This is probably squirrels – they can carry surprisingly large fruit away, but most will be left, with maddening gnawed holes in them. Short of trapping, there's not much you can do to deter squirrels, unless there's an even more inviting crop next door.

Nibbled fruit on the tree is usually due to birds – a number of species, from blackbirds to jays and crows enjoy apples. They like the fruit before it's ripe. If it's a bad problem, netting the tree will avoid bird damage.

Is the fruit being eaten from low-hanging branches and off the ground? If the tree is in a rural location – and without a deer fence – deer are a possibility. Badgers, too, can be fruit thieves: the evidence here will be the scrapes on the tree's bark, left by their powerful claws. If they're a bad enough problem to warrant it, both badgers and deer can be kept out by electric fencing. In both cases, though, they will eat only what they can reach, so unless the tree is a very small one, thefts will be limited.

If ripe fruit is vanishing whole from the tree, don't dismiss the idea of human intervention, especially on open territory such as allotments – do you have apple scrumpers?

Some owners decide to put up with losing a certain percentage of their apples, so long as enough of the crop is left for them. Don't leave the crop on the tree longer than necessary, though – pick it as soon as it is ripe.

When squirrels are the apple thieves, they tend to drop their nibbled treasures on the grass, rather than leaving them on the tree.

Q What does 'heritage' mean?

THERE'S A LOT IN PRINT and online about both 'heritage' and 'heirloom' varieties of fruits and vegetables. What do the terms mean – and, if you choose them, what difference will there be between 'heritage' and ordinary cultivars when it comes to the crop?

It sounds appealing: the gardener gets to grow vegetables that are just like those that their grandparents grew – with the implication that they will be buying a flavour and an authenticity that they wouldn't get from 'ordinary' varieties. And it's true that, having stood the test of time, a heritage cultivar will be good of its type. Realistically, though, most heritage types have changed over the years, and their present qualities will depend on how carefully their breeders have weeded out the more substandard examples over time. If they have been meticulous, then the present-day examples of heritage crops may be even better than the original versions.

A

'Heritage' and 'heirloom' are both terms that are used fairly freely. They indicate that the cultivar – the variety – has been recognised for a long time, typically for 50 years or more.

Open pollinations versus hybridisation

Heritage cultivars are invariably 'open pollinated' – that is, they are pollinated by natural means, including insects, birds, the wind and so on. This means that they have adapted over time to the local growing conditions, and if you collect seeds from them to grow the next year, you're likely to get a

'Heritage' varieties are often more eccentric and irregular in form than their commercially produced cousins.

WHICH IS BEST – HYBRID OR HERITAGE?

It's important to remember why heritage fruit and vegetables exist in the first place – their original growers valued them enough to save the seed. Generally they haven't been developed for large-scale production as hybrids have, but this may sometimes mean that they're more individual. Most of today's gardeners opt to experiment with a mix of hybrids and heritage – and many enjoy collecting the seeds of the latter, so that they can create an entire cycle of seed-to-crop-to-seed for themselves.

similar crop, as the seeds will breed true to type. Note, though, that while all heritage varieties will have been open-pollinated, not all open-pollinated plants are heritage – the heritage part, as the name implies, refers to the length of time for which that specific variety has been recognised.

By contrast, hybrid cultivars (labeled F1 on seed packets) are reproduced by means of controlled pollination – that is, they have been artificially pollinated. Hybrids are commercially cultivated to reproduce specific desired traits in a plant, and their first-generation cross will have a quality called 'hybrid vigour' (heterosis) – that is, it will grow and crop particularly strongly. From the point of view of commercial growers, whose aim is to produce a large number of plants with identical characteristics, this controlled and predictable result is ideal. The disadvantage of hybrids is that subsequent generations will not grow true to type and will invariably be weaker than the first F1 cross. If you grow hybrids, collecting seed and sowing it for the next year's crop isn't an option: you will need to buy new seed annually.

Tomato flowers – or their anthers – only let their pollen go when shaken; bumblebees are efficient helpers here.

Q Can I beat blight?

It's frustrating when your tomato vines flop, or your potatoes rot underground. The most common culprit is tomato or potato blight, a fungal infection that attacks particularly in warm, wet weather. It's spread by spores, carried on the wind. Can you beat it, and grow a healthy harvest?

Late blight, *Phytophthora infestans*, on potato, *Solanum tuberosum*

A

You can choose blight-resistant cultivars and/or early croppers when you're buying seed or seed potatoes. Or, in the case of tomatoes (and if you have access to a greenhouse), grow under glass – where they won't usually suffer from blight.

Try for an early crop

In addition to choosing a resistant cultivar or growing their tomatoes in a greenhouse, some gardeners deliberately pick early croppers – tomatoes that flower and fruit early in the season – on the principle that if they do get blight, it will strike when they're finishing cropping, rather than midway through. It is unusual for early potatoes, too, to suffer from blight.

How you know it's blight

Blight causes the sudden and speedy collapse of both tomato and potato plants – the leaves shrivel and turn brown first, then stems become mushy and watery, and any fruit on the tomato vines rots. With potatoes, the effect on the foliage is similar and the tubers may rot underground – or fail to store when harvested. If you are alert to the symptoms of blight – rapidly browning, shrivelling foliage – and pick off and burn any affected leaves as soon as they appear, sometimes it doesn't take hold. It travels fast, though, and by the time a quarter of the foliage is affected, the plant usually can't be saved.

If you are unlucky enough to have a year in which blight strikes, clear up promptly and carefully, and burn the remains of the affected plants rather than putting them on the compost heap. This is particularly important in the case of potatoes, as leftover potato plants or the potatoes themselves, left in the vegetable garden, may overwinter the spores of the blight, meaning that it can re-emerge next year. If your potato plants show signs of blight, cut and get rid of the foliage immediately, and lift your potatoes within two weeks. Let them dry in the air for a couple of hours before storing.

BLIGHT-RESISTANT TOMATO CULTIVARS

While the majority of early potato cultivars are less blight-prone than later-cropping varieties, some specific tomato cultivars, again mostly early croppers, have been developed to be less susceptible to blight. Try the following varieties for a blight-free harvest.

- **Cordon, or indeterminate tomatoes.** These need to have their side shoots removed from a single stem to crop; they will always need support from canes, string or both, and they can grow very tall – 2.5m or higher. Semi-determinate tomatoes have the same qualities as cordon tomatoes, but don't usually grow so tall.
- **Bush, or determinate tomatoes.** As the name suggests, these are more compact and shorter than cordon types, don't need to have their side shoots removed and will manage with a single cane support.
- **'Berry'.** This type has sweet, juicy heart-shaped cherry tomatoes.
- **'Crimson Crush'.** Trumpeted as the 'world's first 100 percent disease-resistant tomato' when first introduced in 2014, this type bears a generous crop of large fruits weighing up to 200g each.
- **'Ferline'.** This cultivar has medium-sized dark red fruit, with deep flavour. Good all-rounder for both salads and cooking.
- **'Legend'.** This type gives medium to large, almost seedless beefsteak tomatoes – slightly flattened in shape and with excellent flavour.
- **'Red Alert'.** This variety has sparse foliage (reducing humidity around the plant), so it is less likely to attract blight. Its flavour is good and it is notably early to crop.

'Ferline', one of the best all-rounders of all the varieties, has a rich 'tomatoey' flavour that persists when cooked.

Q Can I grow potatoes in shallow soil?

TRADITIONALLY potatoes were planted in a trench and then covered over with soil as they grew, until the developing stems were heaped up with soil. Will they crop in shallower soil, or do they need the depth to produce a crop?

A

Potatoes are an amazingly resilient and tolerant crop, and are great for novice gardeners, as they thrive in most soils and many conditions. You can grow them in shallow soil, in large containers or in open soil under a layer of black plastic.

The conventional way to grow potatoes is to plant them in a shallow trench, which is filled in after planting, then to 'earth up' – to heap soil up around the shoots – as they grow. The tubers form in the resulting ridge of soil.

The most important thing to remember when growing potatoes is that light must not reach the growing tubers. If it does, they turn green and are no longer edible.

Alternative growing methods

If you want to grow potatoes in open shallow soil, you can use plastic rather than earth to keep the light from the developing tubers. You do this by spreading black plastic sheeting over the fertilised, raked and watered plot, weighting down the sides with stones or bricks, then cutting x-shaped holes in the sheet at intervals of about 30cm, tucking the corners back and planting a seed potato in the exposed soil.

Water through the openings in plastic as the plants grow. The plastic keeps the light out, and the tubers grow near to the surface.

Growing potatoes in containers

You can grow potatoes in large containers provided that they have drainage holes. Make some yourself if the container doesn't have them ready-made. Plastic dustbins – ideally with a diameter between 45 and 60cm – work well. Even a heavy-duty plastic compost bag can work if you roll down the top a little (you can roll it up again as you add soil to earth up the potatoes).

You should use good-quality potting compost, and fill the container to within 20cm of the top. As the potato plants grow, add compost to within 5cm of the top. This means that the potato tubers will form in the new compost,

HOW LONG DO POTATOES TAKE TO GROW?

That depends which sort you planted, and when you planted them.

- 'Earlies' are what we call 'new potatoes' – they are usually planted in March, and are ready to harvest from June to July. When the plants start to flower, the potatoes should be ready.
- 'Second earlies' are mid-season potatoes, usually planted early to mid-April, and lifted in July and August. As with earlies, when the plant starts to flower, the potatoes should be ready.
- 'Maincrop' potatoes take longer to form – they're usually planted in mid- to late April and are ready from the end of August to October. If you want to store your maincrop potatoes, wait until the plants' foliage begins to turn yellow, then cut and remove the plants, but leave the potatoes in the ground for another 10 days before lifting them.

To harvest the full crop in one go, dig the plants up – carefully, so as not to cut or bruise the tubers – to reveal the potatoes formed around the roots. If you're growing on a small scale, particularly in containers, and want to harvest bit by bit, you can gently scrape away the top soil with a trowel or your hands and collect a few of the larger tubers, then re-cover and leave the smaller ones to grow.

and there will be enough space at the top of the container to allow you to water effectively.

Three seed potatoes per bag or container will be enough to yield a satisfactory crop. Containers are especially good for growing new potatoes ('earlies') or 'second earlies', both of which grow and crop quickly.

One of the cheapest containers for growing potatoes: a heavyweight plastic sack. Gradually unroll the edge as the plants need earthing up.

How can I grow more vegetables?

WHAT'S THE BEST WAY to get great yields from the vegetables you grow? And are there any sure-fire ways to guarantee bumper crops?

To get great crops, consider the following points:

Sow at the right time – generally, sowing early in a plant's season will lead to a heavier crop. A greenhouse, polytunnel or even a cold frame is helpful to get plants going. If you don't have one of your own, consider begging a corner in someone else's.

Think about whether you would benefit from raised beds, particularly if your soil is clay that drains poorly. They take a little work to make and fill, but it pays off, as not only are they easier to work, but the enriched, deep soil in the beds will encourage a good crop.

Check the spacing of your plants: if you put them slightly closer together (though not to the point of crowding), you'll fit more in and although the yield from individual plants may be slightly lower, the overall yield will be higher. Too close together, though, and the yield may drop overall.

Some climbing crops will often give a great total yield if they're given slightly lower supports – very tall stakes for peas and French beans look impressive, but lower-growing, closer-spaced plants crop better.

Use your space economically – look at doubling-up ('intercropping'), and grow two crops in the same spot – for example, you can sow beetroot between rows of onions before the onions are harvested – the beets will get going, and by the time they need more space, the onions will have been lifted.

Successional sowing – staggering the crop by sowing several sets of seeds a week or two apart – ensures that you spread a crop out across its full season, rather than having a single glut. The latter can look impressive, but is considerably less useful in the kitchen.

A Make sure you're providing all the basics – that you're using enough manure and fertiliser, and watering regularly. If you have acid soil, adding lime will help to keep your soil at maximum productivity.

Beetroots can share growing room with onions to make the most of the space available.

Q How do I pick the right varieties?

THE RANGE of seeds and seedlings offered by seed catalogues and nurseries can seem overwhelming. How can you narrow the choice down to what you want when, with so many appealing options, you can begin to feel that you want everything?

A

Familiarity with seed catalogues and packets will help you make the right choices, if you read them carefully. While flavour is always going to be relatively subjective, most of the other variables – season, height, spread, yield and appearance – are accurately described.

If you find too much choice a problem, start by sticking to one or two catalogues or online sources rather than jumping between several. The RHS Award of Garden Merit (check out the RHS website) is a good place to start – the award-winners have been rigorously tested and re-tested: they not only made the grade when they won the award, but they are re-checked regularly to make sure that they haven't lost their edge.

Look for the qualities you want in a plant's description, but be alert to anything that isn't mentioned. For example, if a tomato is cited as a reliable heavy cropper but the description doesn't mention taste, this may be an indicator that it's not going to be the best choice for flavour.

WILL OLD SEED GROW?

Seed packets always have a sow-before date on them; if you have some stored away, and it's past its sell-by date, will it grow, or should you just throw it away?

Plant it. Opinion on how quickly seed degrades varies – and this will also depend on the kind of seed – but there's a good chance that, in the right conditions, even a year or two after its plant-by date, most of the seed will germinate. In 2012, seeds recovered from a deep-frozen Siberian tomb germinated, and were subsequently carbon-dated. They proved to be more than 31,000 years old.

Q Which vegetables are child's play to grow?

Wild strawberry, *Fragaria vesca*

GETTING CHILDREN INVOLVED in growing their own crops is a great way to start a gardening interest, with the added benefit that even picky children will usually eat vegetables that they've had a hand in growing. So what are the best crops to spark their enthusiasm?

A Children like quick results, so start with a few speedy crops, such as radishes or salads, so that they can see progress. The treasure-hunt aspect of digging up vegetables often appeals, as do crops that can be picked and eaten on site, such as strawberries and peas.

The classic crops to grow with children – the ones that you may well have grown yourself with your parents or grandparents – are radishes, tomatoes (particularly cherry tomatoes that can be picked from the plant and eaten like sweets) and strawberries. But it's easy enough to broaden this repertoire. If you garden alongside them, with plenty of encouragement, children will quickly gain enthusiasm for growing most fruit and vegetables, with the possible exception of some of the larger, more knobbly root crops. Modern varieties of mini-vegetables may have particular appeal: beetroot, finger carrots and even mini-cauliflowers and kale are all available.

A corner of their own

Unless your garden is tiny, try to give children a space of their own, so that they can grow crops with help from you, rather than the other way around. Don't palm them off with the shadiest, driest spot, either – set them up for success with a small plot in a sunny, open corner. If your garden really is

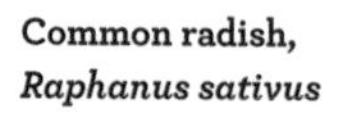

Common radish,
Raphanus sativus

too small, show them how to raise crops in containers: even a five- or six-year old can, with a little help, raise strawberries, some new potatoes or a courgette or squash in containers. Encourage projects that offer a reason to go and check on progress daily, such a growing an extra-big pumpkin for Halloween.

Day-to-day jobs

Involve children in the ordinary work of gardening: weeding, watering, laying on compost and mulch. Gardening is a sensory pleasure, and most children will enjoy the muddy, messy aspect of it. With the under-10s, give them plenty of information, but don't be too much of a perfectionist – concentrate on the 'do's' rather than the 'don'ts'. With slower-growing crops, such as sweetcorn, start a countdown to when cobs will be ready to eat.

Fast-growing and unfussy vegetables, such as courgettes and squashes, offer a speedy pay-off for younger gardeners.

CUCUMBER OR WATERMELON?

One fairly new arrival on the UK vegetable scene is the cucamelon. It looks like an olive-sized watermelon – stripy-green exterior, pinky-red flesh inside – and tastes like a cucumber, and it has instant 'cute' appeal for children.

If you want to try growing a cucamelon and don't have a greenhouse, buy in small plants that have got off to a good start under glass, rather than growing directly from seed.

Q Do fruits and vegetables need ventilation?

ONE OF THE MOST REWARDING THINGS about growing a good crop is having enough to store and use over the next few months. But what's the best way to store your fruit and vegetables so that they won't spoil? And do they always need ventilation, or will some keep without it?

A

Some vegetables – mostly the underground root variety – are self-contained storage organs for their plants and don't need to be pulled up at all: they can overwinter in the ground until you want to use them. Many others need storing carefully if they are to keep, and may need good ventilation around them to ensure they don't rot.

Some fruits and vegetables keep better than others, but making sure that you're keeping them in the right conditions will help ensure you can continue to enjoy your crop after harvest. Storage conditions must always be dry, and some degree of ventilation is useful except in the case of root vegetables, which can be prone to dry out.

Check your stores

Any fruits and vegetables you're storing should be checked every day or two. If one piece begins to rot, the problem can quickly spread to the rest of the crop, so make a quick daily inspection and if you find any with soft spots or the beginnings of rot or mould, remove these ones straight away.

Carrrots, cleaned and stored in a box of slightly damp sand, will keep for some months if kept somewhere cool but frost-proof.

HOW TO STORE FRUITS AND VEGETABLES

Apples, pears and quince

These need to be kept somewhere cool, dry, dark, well-ventilated – and ideally mouse-proof. Slatted racks or trays are ideal for storage, and the fruit should be laid out, without touching, in a single layer. If you're lucky enough to have a quince crop, store it well away from anything else; the smell is delicious but pervasive and may affect the taste of other kinds of fruit.

Tall wooden racks with open slats are ideal for storing apples. The fruits should be laid out on a single layer.

Stored fruit should be checked daily, in particular pears, which have a small window between ripe and rot. At the first sign of ripening, either eat them or remove them to the fridge where the ripening period can be prolonged for a day or two.

Pumpkins and squash

Harvest them with as long a stem as possible (although don't then handle them by the stem), then leave them to 'cure' for 10 days, either out of doors – protected from frost with fleece or cardboard – or in a greenhouse. This completes the ripening and hardens the skin in preparation for storage. Afterwards, they can be stored in a cool, well-ventilated place such as a shed or cellar. The stored fruits shouldn't be allowed to touch one another, and will keep for up to six months.

Root vegetables

Root vegetables, including beetroot, carrots and celeriac, are usually best left to overwinter in the ground, and then they can be lifted as they are needed. If your soil is very wet and poor-draining, or if there is a likelihood of heavy frosts, lift them and store in a cool place – again, a shed or cellar is fine – in boxes, layered with sand. Root vegetables are prone to dry out and don't need ventilation, although they need to be kept away from damp.

Q Which crops can I freeze?

IF YOU HAVE SPACE in the freezer, this may seem to be an easier option than storage for both fruit and vegetables.

A Most crops can be frozen in one form or another. Knowing what works best for different kinds will ensure your fruit and vegetables are still at their best when defrosted.

The effectiveness of freezing crops varies, depending on the type of fruit or vegetable.

Soft and stone fruit

Fruits such as raspberries, strawberries and currants should be spread out on trays in a single layer and frozen before being tipped into bags, labelled and returned to the freezer. This avoids them turning into solid bricks. With stone fruit, cut them in half and remove the stones before freezing.

Legumes

Peas and broad beans freeze well if you have a glut, but are also among the best commercially available frozen vegetables, so aim to eat them fresh if your crop isn't huge. Other beans such as borlotti also freeze successfully. French beans aren't usually as delicious frozen, but the immature bean seeds are good if podded, so consider setting some aside for the freezer – they taste better than the dried ones if you add them to winter soups and stews.

A FEW TIPS

Some vegetables won't freeze successfully: don't try it with salad greens, chicory or Jerusalem artichokes. You can freeze celery – which is good as a base for stews, soups and stock – but chop it into small pieces first.

Blanching

Blanching with a brief immersion in boiling water stops the enzymes that, after harvesting, convert starch to sugar and dilute the 'just picked' flavour. This process will also take place – very slowly – in vegetables that are frozen. Blanching also helps vegetables to keep their colour.
Cut the vegetables into manageable pieces then boil in a large pan for around a minute. Drain, then freeze spaced out on trays before bagging, labelling and replacing in the freezer. (If you skip blanching, the vegetables will still be fine to eat, but the flavour and colour will both 'fade' over a prolonged period in the freezer.)

Which are the best winter crops?

When a new gardener begins to consider which vegetables to grow, they tend to think about typical summer crops. But to keep the growing season as long and as productive as possible, you need to take winter crops into consideration.

Bear in mind that winter crops often need to be started off in late spring, planted out in summer and can occupy ground for nine months, so if space is short, make your choices carefully. The real advantage that winter vegetables have is that many can 'stand' in the ground until you are ready to eat them, even though growth slows or stalls in the cold season, allowing you to harvest fresh vegetables in the depths of winter.

Salads can be sown from late summer until early autumn. The first sowings will mature before the colder weather, but later ones will probably need protection under cloches or fleece, or in a greenhouse, as the weather gets colder later in the season.

FILLING THE HUNGRY GAP

'The hungry gap' is the time in the garden at the tail end of winter through to spring – from the end of March to May in the UK – when late crops are over and new ones aren't yet ready to harvest. Overwintered cauliflower, leeks and sprouting broccoli are just some of the crops that can be grown to bridge this lean time – check catalogues for late-maturing varieties that will work on this timetable.

Hardy winter green vegetables include sprouting broccoli, kale, leeks, cabbages, parsnips and Brussels sprouts. If you can offer some cover, you can also grow salad leaves, such as rocket, mizuna and mustard greens.

Broccoli 'Purple Sprouting', *Brassica oleracea* (Italica Group) 'Purple Sprouting'

Q Does anything scare off wasps?

MOST GARDENERS WELCOME BEES with open arms. But wasps, their not-so-furry or friendly cousins, are usually looked on much less warmly. Is there any natural way that you can deter them from hanging around while you're tending your garden?

Common wasp, *Vespula vulgaris*

A There's no scientifically proven wasp repellent, and ideally wasps should be tolerated – not only are they pollinators, but also voracious carnivores that eat up many garden pests.

The case for wasps

Wasps play an important role in the garden environment. It is usually only for a brief time in late summer, before they are killed off by the cold, that wasps get dozy and become a nuisance to the gardener.

Existing 'repellents'

There is a long list of things that are held to repel wasps, but research indicates that few of them are really effective. Traditional wasp traps (glass vessels baited with a sweet solution which drowns any unwary insects that crawl in) kill some wasps but research has shown that they seem to attract just as many to buzz around in the vicinity – if anything, increasing the nuisance. The traps are also indiscriminate, and will catch and kill other valuable insect pollinators.

▼ In folklore, it is held that mint repels wasps – but there's no scientific evidence to support this traditional story.

Artificial wasp nests can also be bought – the idea is that wasps, being highly territorial, will steer clear of territory that appears to be taken already. There seems to be no research to back this up.

In folklore, some plants are held to ward off wasps – again, their effects aren't really proven, but if you want to try experimenting with creating a wasp-free zone, try out planting some mint, eucalyptus, citronella and artemisia.

In conclusion, it's best to try to live alongside your wasps – they do far more good than harm.

THE MOLECULAR PICTURE

There is plenty of research into wasp repellents going on. Studies in both the USA and Belgium have concentrated on whether or not certain essential oils will deter them. Experiments under controlled conditions have found that citronella, clove and geranium oils have a deterrent effect. Alas, the conclusion is that they are, as yet, unlikely to work in the less controlled environment of the garden – the scents dissipate too quickly in hot weather and blow away if there is a breeze.

However, Belgian scientists have carried studies a step further: they're investigating the molecules within the essential oils, and may yet be able to isolate one that will act as a super-deterrent. It's not yet known how the molecules work; they are believed to mimic the chemical signs that insects use to communicate, so an effective molecule might be used to jam the wasps' communications.

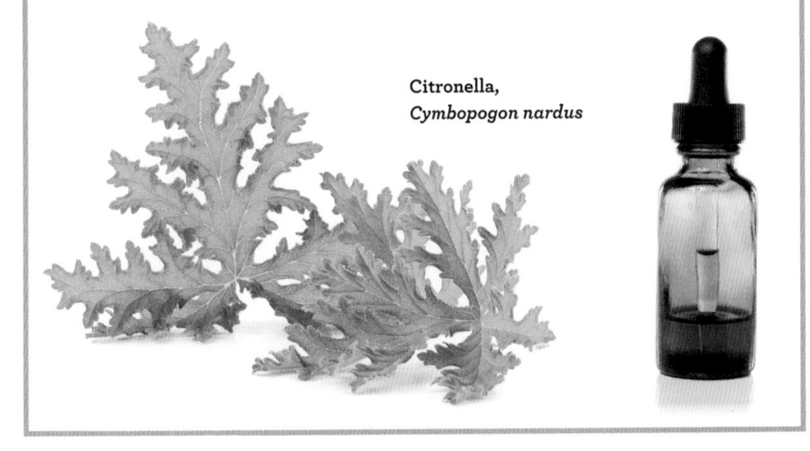

Citronella,
Cymbopogon nardus

Q Can I grow crops in containers?

WHICH CROPS can be grown in containers? Can the repertoire be extended beyond tomatoes, herbs and potatoes?

With many gardeners having limited outdoor space – in some cases just a small yard or patio, or even a balcony – container gardening is as popular as it has ever been.

Choosing the right container

It's important that the containers you use are large enough. The crop's roots need space to grow and there must be enough soil to offer sufficient nutrients for the plant to grow healthily. Herbs can often be grown in smaller pots, but containers for vegetables should be at least 45cm in both depth and diameter. The best all-round growing medium is soil-based potting compost.

Choosing your crops

Don't be afraid to experiment – it can take quite a lot of trial and error to find the system and crops that work best for you, and in the meantime you can have fun trying new varieties.

Rather than choosing each variety singly, plan ahead. Containers can be used to grow a succession of crops.

For example: carrots can be sown in late February and harvested in June. French beans can then be grown in the same container and harvested in September. Spinach can be sown in February, and will run to seed by June. Replace the spinach with courgettes, and they will produce by the end of August through September. Fast-growing mini lettuces, such as Tom Thumb or Little Gem, can be sown in January, harvested in June and replaced by leeks for a winter harvest.

The case for growing bags

Growing bags don't look particularly decorative, but they can be an efficient and cheap way to grow crops, such as salad greens, courgettes, peppers, cucumbers and tomatoes, in a small space. If you don't like how they look, you can empty them and rehouse the plants in a similarly sized container, or mask them behind other containers. Always shake the bag thoroughly, to loosen the soil up, before opening.

A Almost all vegetables, from beetroot to broad beans and spinach to cucumbers, can be grown in containers, provided that the container is large enough, that it is planted up with the right soil or compost and that watering and feeding happen regularly.

Q Do I need a cold frame?

WE'VE ALL ADMIRED the long rows of beautifully made cold frames ranged along the lower outer walls of greenhouses in Victorian kitchen gardens, but do they have a place in a small, modern garden? What can you use a cold frame for?

A cold frame isn't an absolute necessity in the garden, but it is incredibly useful for all sorts of jobs, from hardening off seedlings to overwintering bulbs. Once you have one, you really will wonder what you did without it.

Ready-made cold frames are convenient but expensive. Home-made is a good option if possible. It is easiest to make one with an old glazed window as the top, and trimmed scrap wood for the sides. The ready-made variety tend to be either glass or plastic on both top and sides. The lid needs to be easy to open and to prop securely while you're taking things in and out of the frame.

Generally, cold frames with a lower, flatter structure are preferable to the taller ones with shelves – the larger floor area is more useful.

To go one step up, try a mini-greenhouse or lean-to structure. These are handy, and will allow you to grow crops that benefit from being under glass like tomatoes or aubergines.

FIVE USES FOR A COLD FRAME

- For housing seeds, such as hellebores, that need a winter chill to germinate but that don't appreciate winter rain.
- For hardening off seedlings, and to overwinter both seedlings and cuttings – for example (vegetables) lettuce, broad beans, cauliflowers, broad beans, (flowers) hardy annuals and perennials.
- For growing young plants before they're planted out.
- For giving extra warmth to low summer crops such as cucumbers and melons.
- For striking cuttings.

Q What's making holes in my potatoes?

HOLEY POTATOES are a common problem, but identifying the culprits and – most importantly – stopping their attack, can be a challenge. Check out a line-up of the usual suspects and learn how to beat them.

A

Slugs love potatoes and are by far the most likely offenders. Another less common possibility is wireworm. There are also freeloaders to consider, including woodlice, large nematodes and millipedes, who don't start the damage themselves, but take advantage of it.

Slugs

The good news is that if, as is most likely, the problem is being caused by slugs, you can use nematodes against them very effectively. The nematode mix is added to water and watered on to the soil in spring, as soon as the temperature is high enough – the soil needs to be warm and moist for the nematodes to get to work. The microscopic creatures will enter the slugs' bodies and deliver lethal bacteria that will kill them from the inside out.

Other offenders

If you treat for slugs, millipedes and centipedes won't usually cause serious damage, and although you may see woodlice on holey potatoes, they don't start the damage off – they are simply following the slugs' lead.

Wireworms, however, cause very recognisable damage in potatoes – they are the orange-brown larvae of the click beetle, between 12 and 25mm long, and they burrow into tubers leaving long, thin tunnels, quite unlike the big, soggy cavities left by slugs. Wireworms are only present in very weedy, overgrown gardens or in a

Green cellar slug
Limacus maculatus

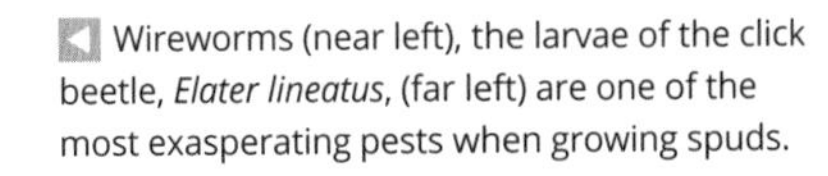

Wireworms (near left), the larvae of the click beetle, *Elater lineatus*, (far left) are one of the most exasperating pests when growing spuds.

newly dug potato bed that has until recently been covered with grass. If the bed is dug well before the potatoes are put in, the larvae may be seen (and if you do see them, you should remove and squash them). Nematodes that act specifically – and effectively – against wireworms are also available.

NEMATODES AND HOW THEY WORK

If you've only ever read about nematodes being used against garden pests, you may be under the impression that these microscopic worms are always on the gardener's side. But every square inch of soil in your garden is home to hundreds, or more likely thousands, of nematodes, and each of the 25,000 or so known species is different. Some live independently, feeding on all kinds of matter, from algae to fungi, while others live as parasites.

On the one hand, gardeners may use nematodes as an efficient and environmentally friendly control against the slugs that are eating their potatoes – but on the other, one of the most damaging potato pests of all is a nematode – it's the potato cyst nematode, which ruins both domestic and commercial potato crops all over the world.

The specific nematode that kills slugs has the less-than-catchy name of *Phasmarhabditis hermaphrodita*. Although it was only developed and marketed as a biological control in 1994, it had been identified in 1859 and long known to kill gastropods. Its methods are gruesome (stop reading if you're squeamish): after entering the slug's body, it releases bacteria that kill its host, then goes on to use the corpse as a reproduction chamber.

Q Will herbs put off pests?

TRADITIONAL GARDENING LORE has long held that many plants, and in particular herbs, will help to repel various pests from the vegetable garden. How true is this, and which pests will be affected?

A

The jury is out on the pest-repellent qualities of herbs, which haven't necessarily been borne out by research. However, what has already been proved is that most flowering herbs are excellent for attracting insects that are good for the garden – both pollinating plants and feeding on common garden pests.

It's hard to reproduce real-life conditions in controlled tests. Experiments have proved that the aromatic molecules that give herbs their characteristic scent and flavour have an effect on insects, either attracting or repelling them; out in the field, though, the results are less clear-cut. However, there's no harm in experimenting with some of the herbs that allegedly repel specific pests – if they work, it's a win-win in the vegetable garden.

HOW DO PESTS CHOOSE THEIR PLANTS?

All insect pests have their preferred host plant variety or varieties. A study at Warwick University indicated that the volatile oils that give herbs their scent didn't really seem to discourage insect pests from landing on the leaves – they weren't driven off by the smell. On an area with a wide variety of plants, the insects seemed to choose any green area to land on, and only after landing to decide whether it was their host plant (in which case they'd stay) or a non-host type, in which case they'd stay for a short time and then fly off. They didn't land on bare earth, but different areas of green seemed to attract them equally – even when these were actually artificial plants. The whole process, it seemed, was more hit-and-miss than the researchers had expected. The results are far from certain, but it may be that what companion plants really do best is to waste insect pests' time and distract them from their intended targets.

WHICH HERB FOR WHICH PEST?

If you'd like to try them out, here are some herbs with the pests they are alleged to deter:

Chives and coriander for aphids

Oregano, thyme, rosemary and mint for cabbage moths

Rosemary and sage for carrot flies

Southernwood for fruit tree moths

Marigold for asparagus beetles

Mint and catmint for flea beetles

Should I soak seeds?

YOU MAY HAVE HEARD recommendations that hard seeds should be given a soaking before sowing. Is this necessary and will it help to speed up germination?

There is no one-size-fits-all method when it comes to sowing seeds; they vary hugely, from large with a tough outer coating, to microscopic. It's the larger, harder seeds, like those of lupins, that may need nicking (see below), and some seeds may germinate faster if they're soaked before sowing.

Some seeds, particularly those from trees and shrubs, may be soaked before planting to soften the outer coating and decrease the amount of time the seeds take to germinate. For many seeds, an increased moisture level is an indicator that it's time to grow.

When it's appropriate, seeds are usually soaked for between 12 and 24 hours. If they are soaked for too long, it may make them susceptible to rot, so it's important not to overdo it, and always refer to instructions for the specific seed type.

If a seed has a very hard outer coat, it can be nicked ('chipped') with a sharp, clean knife to help germination. Seeds that benefit from nicking include sweet pea, nasturtium, morning glory and lupins.

HOW LONG DO SEEDS TAKE TO GERMINATE?

This is a true how-long-is-a-piece-of-string question. Temperature is the most important factor: for example, in warm soil, a carrot seed may emerge in just five days, while under cold conditions, it may take as long as 20 days to germinate. Given favourable conditions, the vast majority of seeds take 90 days or less – often fewer than 30 – to germinate. A very few, though, can take a year or more to get going, and some need extremely specific conditions to germinate, such as two successive chilly periods. Every plant has evolved to make the most of its chances of survival, and making seeds that will germinate only when the conditions are as close to ideal as possible is part of that evolution.

Q How many seeds should I sow?

WHAT RATE OF RETURN is it reasonable to expect from a packet of seed? Can you rely on most of the seeds germinating?

A Some packets contain so many seeds that it's tempting to sow sparingly and save some for the next season. But it's usually a good idea to be generous. The success rate also depends on whether you are sowing indoors or out.

Seed viability is legally specified in many countries. In the UK, most vegetable seeds must be certified 55 percent viable – but this isn't necessarily an accurate guide to the number of seeds that will germinate and go on to become viable seedlings. All sorts of variables can affect how many will actually grow, from sowing conditions to the quality of the seed itself. A seed is still a natural product, however strict the conditions were during its production.

Seeds are also sold with the expectation that they will be used fresh – and usually, the fresher the seed, the higher its likely viability.

So what are the best broad guidelines?

Valuable seeds are best sown indoors – where growing conditions can be controlled and so the chances of germination are improved – since there are just a few seeds in each packet and you're unlikely to have 'spares'.

With easy-going crops that can be transplanted, sow thinly to avoid the excessive thinning of seedlings later but sow a short length of soil more thickly at the end of each row, to give some surplus seedlings to fill any gaps.

If you're sowing fussier outdoor crops that don't like to be moved, such as carrots and parsnips, try 'station sowing'. This means planting between three and five seeds at each point where you ultimately want just one plant. Although it will mean thinning where excess seeds have germinated, the pay-off is an even row of plants without wasteful gaps.

If you find you've generated a large quantity of seedlings and end up wasting many when planting and thinning out, harden your heart: seeds aren't usually very costly, and seedling wastage is a natural part of the gardening cycle.

Why won't my apple tree crop?

APPLE TREES are usually prolific croppers, so it is frustrating when one fails to produce a satisfying bounty, particularly if it has cropped well previously. The possible reasons are numerous, but many can be fixed for next year.

Apple tree blossom is fragile, and sadly a late frost may damage it to the point where it's no longer viable.

Run through the questions on this page and ask yourself which apply to your tree, then see what you can do to fix the problem.

Questions to ask

Is the tree old enough?
Some young fruit trees can take a number of years to mature and bear fruit.

Was the tree pollinated?
Poor weather can mean that pollinating insects were in short supply when the tree needed pollinating. Most apples also need another tree of a different cultivar to act as a partner in pollination. Gardeners should take claims that a cultivar is self-fertile (often made at point of sale) with a pinch of salt.

Did the tree flower?
Without flowers, there won't be any fruit. The flower buds may have been eaten by birds – chaffinches, for example, are partial to them – or by pests: if you're unlucky, apple blossom weevil can decimate a crop.

Were the flower buds damaged?
Late frosts can damage blossom beyond repair. Smaller trees can be protected with fleece, and some late-blossoming cultivars avoid the risk of frost damage.

Was the tree pruned?
Poor pruning can damage shoots that would have borne blossom in due course.

Is your tree all-round happy?
Does it get enough sun? Does it get enough water? Is it growing in fertile ground?

Biennial croppers

If you had a good harvest last year, your tree may be one of the biennial croppers, which alternate a good year with a poor one. Although this can be a natural tendency, it can be countered by bravely pruning out much of the blossom at flowering time during an 'on' year.

WHY TREES NEED PRUNING

As no tree or bush ever looks like an illustration in a book, we've avoided a pruning guide here, and looked instead at how pruning works. Seek out online videos if you want to prune a tree and are uncertain where to start, and refer to the RHS site (rhs.org.uk), which gives clear guidance.

- Trees have a natural propensity to flower and fruit, so if you left them to their own devices you would still get a crop, although it might not be a very good one, and would ultimately come from rather large trees.
- Light pruning results in many small shoots, with little shoots called spurs that will carry most of the crop. Heavy pruning is generally less desirable, as the tree fights back and it tends to produce excessive unfruitful regrowth.
- Relatively few flowers need to be pollinated to produce a generous crop, because apples and pears are large fruits. Removing some flower buds will not be disastrous for the crop – in fact, if the tree finds itself overloaded as the fruit develops, it may drop some naturally in order that the remainder can grow to full size (see 'the June Drop', p74).
- Good pruning deters congestion in the middle of the tree, strong vertical shoots and weak or diseased growth. The desired result is a tree with shoots that are about 45cm apart at their bases, and with a slight outward lean. The fruit should then pull them downwards and outwards, keeping the centre of the tree open.

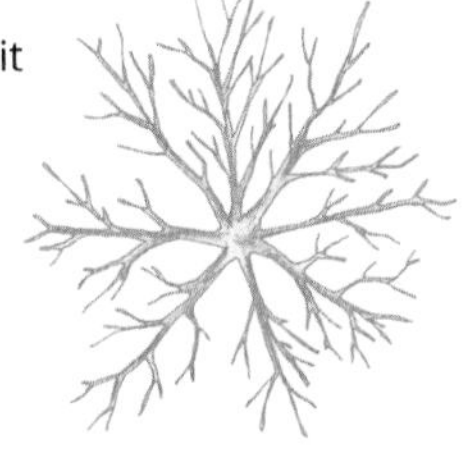

▲ Good pruning opens up the centre of the tree (top). Too many twigs left competing can result in congested growth (bottom).

Chapter 3

On the Ground

Why is my lawn lumpy?

A LUMPY LAWN is a question of perception. It may seem that there are lumps, but far more likely is that there are hollows making the remaining lawn appear to be lumpy. Whatever the cause, an uneven lawn can be unsightly and make mowing difficult.

Uneven wear of a lawn can form hollows, particularly in gardens used by children, or along 'desire lines' (favoured short-cuts). Animal activity under the soil – moles and voles in particular – can also cause the lawn to sink as their tunnels beneath collapse. Gradual erosion of soil but not underground stones and rubble (common in gardens of newly built houses) can also create lumps.

Tree roots

Large trees can cause lawns to sink. When they're still standing, their roots can remove enough water from an area to cause hollows. Even when a substantial tree has been felled, the old roots, left in the soil, will eventually rot, leaving hollows for the lawn to sink into.

Levelling lawns

Redressing a lumpy lawn can be the work of an afternoon or of a season. In the longer term, minor hollows can be raised to the level of the soil around them with the addition of fine topsoil and grass seed. This process

> **A** Attaining a bowling-green lawn is a question of maintenance. You will need to rake it after mowing, and sprinkle topsoil and grass seed to fill hollows.

▼ Moles aren't just responsible for molehills; they may also cause areas of lawn to sink, as soil collapses into the tunnels they make.

RAKING IT OVER

Two main causes of lumps in the lawn are worm casts and an uneven build-up of 'thatch' (old grass clippings, moss and other debris). In autumn, vigorous raking, known as scarification, will help to spread the casts and thatch evenly, or remove them altogether.

Raking the lawn is best done seasonally: give your grass a light rake in spring, then don't rake at all in summer, saving the vigorous scarification for autumn.

can be repeated as necessary until the lawn is renewed. Annual aeration of compacted areas – using a garden fork to create lots of little holes in the ground – will also aid new root growth.

Lumps can be lowered and hollows raised quickly by removing or adding topsoil beneath the turf layer. Using a sharp spade or half-moon, cut a 'T' shape into the centre of the lump or hollow and slice underneath the turf until it can be peeled back. Add fresh topsoil to hollows, or dig out soil (or stones if they are found) from lumps and replace the turf, making sure to tread firmly on the turf to ensure it has a good connection with the soil. Fill in any cracks with topsoil and water well.

A flat lawn can be achieved through regular maintenance. Clear away old clippings and add topsoil to hollows.

Q What can I do with dry soil?

IN THE PAST, dry gardens were viewed as challenges to change: the plan was usually to enrich the soil first, and then look at possibilities for artificial irrigation. Today, though, owners of naturally dry gardens are more likely to make the most of what they already have.

The work of Beth Chatto, the well-known British gardener, did a lot in the 1970s and 1980s to raise awareness of just how good dry gardens can look. Her 'dry' gravel garden in Essex, southeast England, which is famously never watered, became ubiquitous in gardening features and magazines. Gravel gardens have since become popular where the conditions are dry but sunny. They have several advantages: once set up, they're low-maintenance, they are popular with wildlife and they work very well with Mediterranean planting schemes. They can look wonderful in small spaces, or filling an area within a larger garden.

A Most gardeners have become increasingly keen to work with the conditions that their gardens give them naturally – and if these include dry, free-draining soil there are still plenty of plants that will grow there happily.

There are two options when making a gravel garden – you can put down landscaping fabric and make slits to plant through before laying down the gravel over the top, or you can put the gravel directly onto the prepared soil. Either way, the soil should be thoroughly weeded and prepared first. The advantage of the landscaping fabric is that your gravel garden won't have weeds – gravel laid straight onto soil will mean that a certain amount of weeding will still be necessary.

◀ The African daisy, *Osteospermum*, is a reliable performer in a dry, sunny spot – and is available in a range of vivid pinks as well as cream, white and yellow.

FIVE PLANTS THAT DO WELL IN DRY, SUNNY CONDITIONS

Plants that have evolved to cope with dry soil tend to have foliage that has adapted to minimise the amount of water lost from them – which can include very narrow leaves (such as lavender) or fleshy, waxy leaves (such as African daisies, *Osteospermum*). In addition, they may be especially deep-rooted or have fleshy roots that store water.

Lavender, *Lavandula*. The narrow grey leaves of lavender plants don't lose much water, and it also contain volatile oils, which evaporate and help to cool the foliage.

Spurges, *Euphorbia*. Particularly suitable are the vivid-yellow *E. cyparissias* (which has vivid yellow flowers) and *E. myrsinites* (lime-green flowers). Many euphorbias have grey foliage, and fleshy roots that help them to store what water is available.

Evening primroses, *Oenothera*. Both the yellow- and pink-flowered forms love dry soil – although be aware that, lovely as they are, they can sometimes run wild if they find a particularly hospitable spot.

African daisies, *Osteospermum*. Fleshy, waxy leaves help to reduce the amount of water that evaporates from these plants.

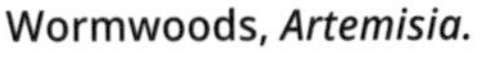

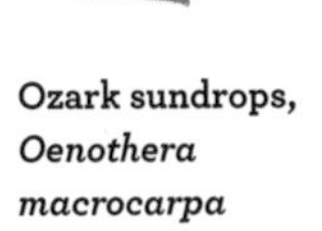

Wormwoods, *Artemisia*. A large family of fragrant plants that is mainly grown for their feathery grey-green foliage. Two of the most popular varieties are 'Powis Castle' and 'Lambrook Silver'. Both have small yellow flowers.

Ozark sundrops, *Oenothera macrocarpa*

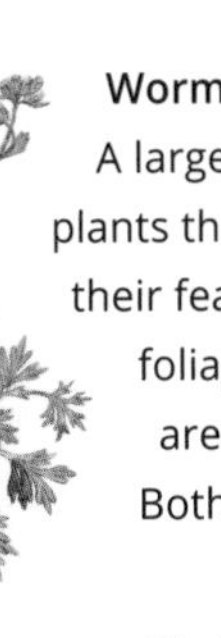

Glacier wormwood, *Artemisia glacialis*

Q What can I do with soggy soil?

GARDENERS USUALLY WELCOME RAIN, but what if your garden's problem is too much water? Although gardens are seldom all-round, year-round wet, many can become waterlogged after heavy rain in summer, as well as in winter. What's the best way to cope with a wet garden?

A As with dry gardens, the best way to deal with a wet site is to work with what you've got. Test how wet or compacted your soil is, then plan and plant accordingly. Planting in raised beds or on low mounds of soil can be helpful when you want to grow plants that won't tolerate prolonged soaking.

To see how wet and/or compacted your soil is, dig a hole around 60cm deep in a 'typical' spot, water the hole, cover it with weighted plastic to stop it filling if it rains, and leave it for a day. If, when you remove the plastic, there's still water at the bottom of the hole, you have poor-draining soil.

The problem with poor drainage is that roots need oxygen, which isn't present in waterlogged soil. Lots of organic matter, whether dug in or used generously as a mulch, will help to aerate the soil, but unless your garden is very small, it may be difficult to add enough. (And, although you still often see it recommended, the addition of grit or sand to the soil tends to be less effective, as it's usually impossible to add enough to make a real difference.) Neither approach will help much if there is an overall drainage problem.

Be pragmatic and use a mixed strategy. Build a couple of raised beds filled with plenty of organic matter to create sites where conditions can be controlled. Outside of raised beds, improve drainage around a plant's roots by growing plants in slightly raised mounds of soil. And in other areas of the garden, choose plants that are content in wet conditions.

Common dogwood,
Cornus sanguinea

SIX OPTIONS FOR WET SOIL

Many plants can cope with damp soil with limited oxygen supplies, so this is just a small selection of some of the most appealing options.

Dogwoods, *Cornus*. These are attractive shrubs – particularly *Cornus sanguinea*, which offers white flowers, red stems in winter and deep-red foliage in autumn. The berries are also very popular with birds.

Astilbes, *Astilbe*. Easy-to-grow perennials with fluffy flowerheads that are held above the foliage, and come in a range of pinks, reds and white, and a variety of heights, from 25cm to over 1m. They're also happy in shade.

Sea buckthorn, *Hippophae rhamnoides*. This is a deciduous thorny shrub with silvery leaves, and bright-orange berries on the female plants (you'll need both male and female to get a good crop). The berries are sweet-sour, especially high in vitamin C and popular treats for birds and, in jellies and jams, for humans.

Siberian iris, *Iris sibirica*. This plant's bright violet-blue flowers appear in early summer. The rhizomes spread when it's in a spot that suits it.

Bergamot, *Monarda didyma*. Also known as bee balm, this is a fragrant, deciduous perennial that can grow to 1.2m. The flowers are red, pink and purple, and hybrids are also available.

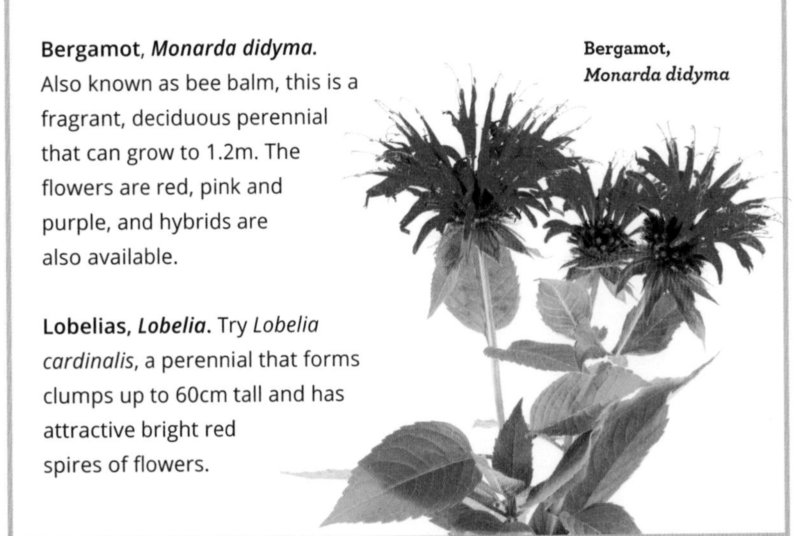

Bergamot, *Monarda didyma*

Lobelias, *Lobelia*. Try *Lobelia cardinalis*, a perennial that forms clumps up to 60cm tall and has attractive bright red spires of flowers.

Q Can I second-guess squirrels?

MANY GARDENERS, especially those in semi-rural settings, would not place grey squirrels far behind slugs when it comes to garden pests. They may not provoke the visceral distaste that slugs do – in fact, they're comparatively cute-looking – but their reputation of being 'rats with furry tails' wasn't earned for nothing.

A

As you're unlikely to be able to rid your garden of squirrels altogether, the better tactic is to find ways of limiting the damage they do.

However engaging it looks, this squirrel can cause an irritating amount of damage in the garden, from digging up bulbs to stealing fruit and nuts.

Grey squirrels were brought to Britain from America – a banker named Thomas Brocklehurst first brought a pair over the Atlantic and released them in his Cheshire grounds in 1876. Other tourists brought more squirrels, which were (briefly) perceived as an engaging

SQUIRRELS AS TENANTS

If squirrels move into your house, it's time to call pest control. Once they get into a roof or loft (usually going in under the eaves, where they can wriggle through surprisingly small gaps) they can damage wiring, insulation and internal timbers, so call the experts and let them deal with the problem. And make sure that any means of entry has been securely sealed when your unwanted tenants have been ejected.

novelty, and before long the native, and much shyer, red squirrel population was strongly impacted by disease the greys brought with them. There is currently an estimated population of between 2 and 3 million grey squirrels in the UK – and they're unpopular with many gardeners.

Repeat offences

Squirrel crimes include gnawing tree bark to get at the tree's sap and stripping trees and shrubs of their flower buds or edible crops, such as pears, apples and nuts. They also make holes in the lawn to bury their finds, and dig up and eat bulbs and tubers, as well as nibbling other crops such as strawberries and sweetcorn.

Fighting back

How can you thwart them in some of their activities? The jury is out on whether or not the squirrel repellers that emit ultrasonic sound actually work. They have the advantage that they're not hugely expensive, and they can be moved around different parts of a larger garden; some gardeners report that the use of these devices has resulted in fewer squirrels – and less damage – but others haven't seen much effect.

The most practical way to guard plants is to protect them with a physical barrier. Newly planted bulbs, strawberries or particularly precious plants can be covered with wire mesh (it must be wire, as squirrels can easily bite through plastic). Mesh can be bent into tunnel shapes – with closed ends – and pegged down.

If you spot any bite marks on the trunk of a tree, act quickly to protect it; 'ring-barking' – when the bark is stripped off around the trunk in a complete circle – can kill the tree. Depending on how high up the damage is, a protective collar of wire mesh may protect it.

If your hazelnuts are split in half, squirrels are the culprit; holes in the shells mean it's more likely to be mice or voles.

Wire netting will keep squirrels off valued plants and crops; the green, plastic-coated option will be less visible against foliage.

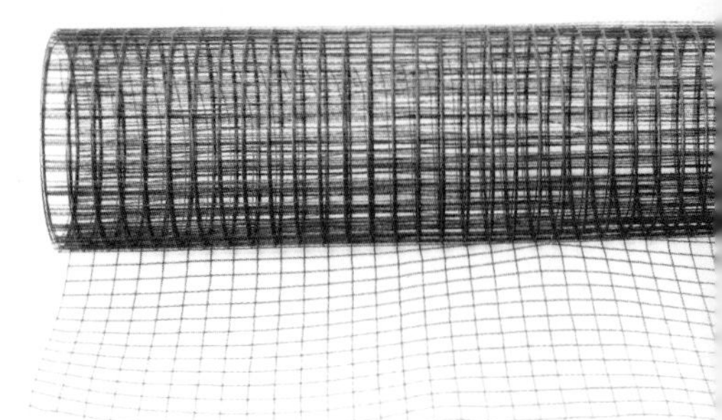

How do I combat cats?

WHETHER YOU'RE A CAT LOVER OR NOT, cats can be irritating in the garden. They sunbathe on treasured plants, they sharpen their claws on the bark of trees and shrubs and, probably the most common complaint, they treat flower beds as lavatories. Wildlife lovers also deplore cats' – very successful – hunting habits.

Even if you don't have a cat yourself, it's not really possible to keep cats entirely out of your garden, but there are a number of ways you can make their visits less frequent.

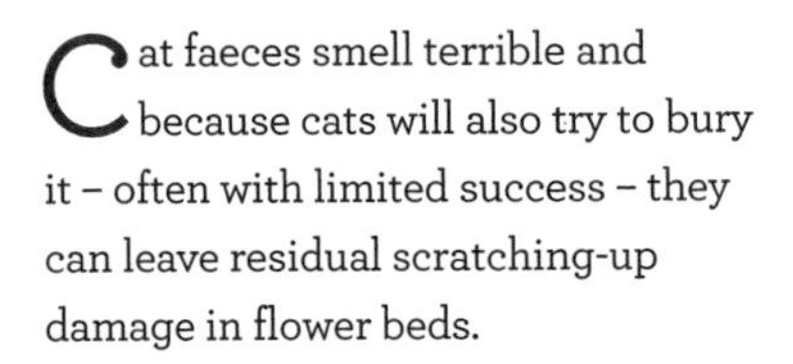

Cat faeces smell terrible and because cats will also try to bury it – often with limited success – they can leave residual scratching-up damage in flower beds.

Make your garden less inviting

Discourage cats by keeping your planting as dense as possible: they prefer to defecate on bare earth, so don't leave any spaces. They don't like damp soil, so keeping flower and vegetable beds well watered will also help to dissuade them. If you don't want an area of freshly sown ground disturbed and you get a lot of feline visitors, it can be protected with plastic netting, pegged down at the margins. Particularly treasured seedlings or young plants, which unavoidably have some bare soil around them, can be given small temporary stockades of cocktail sticks placed upright in the earth around them: cats won't want to walk, or squat, on them.

The battle against cats in the garden can feel like a minor form of guerrilla warfare, especially in cities.

THINGS CATS HATE

There are a variety of options when it comes to substances that cats are alleged not to like, some more outlandish – and expensive – than others. Reports of how successful they are vary, so if you want to try them, start with the cheaper options and work your way up.

▲ Cats dislike ordinary black pepper – it gets in their noses and makes them sneeze.

- Pepper powder, aluminium ammonium sulphate and methyl-nonyl-ketone are all found in a variety of commercial sprays and powders (some owners use neat chilli powder). They're not rainproof, though, so applications have to be frequent. Occasionally, odder substances, such as dried lion faeces, are advertised as cat repellents, but while they might make an amusing present, they're unlikely to give the local cats much of a scare.
- *Plectranthus ornatus* (sometimes sold as *Coleus canina* 'Scaredy Cat') is a plant that produces a horrible smell when it's brushed against, and some cats strongly dislike it.
- You can buy ultrasonic devices that are triggered by movement, which claim to repel cats using a sound outside the human hearing range. They work best in gardens with a fair amount of open space. Again, some gardeners find these devices successful; others don't.
- Directional water sprayers are also available; triggered by sensors, they're intended to soak the cat as it approaches.

Plectranthus ornatus,
Plectranthus ornatus

A completely different approach is to create a 'cat corner' in your garden: plant up a sunny corner with a large patch of catmint, in the hope that it will act as a distraction for visiting cats and that what time they spend in your garden, they will spend there.

Q Can anything help bees?

INSECTS OVERALL, including vital pollinators, are believed to be in decline. It's been known for a long time that bee populations are under threat from a variety of factors, from pesticides used both domestically and agriculturally, to the varroa mite that has been such a problem to cultivated honeybee colonies. As a small-scale gardener, can you do anything to help?

A With many previously wild corners of the countryside having been turned over to agriculture, gardens are now increasingly important havens for all pollinators. Bee-conscious gardening can certainly help not only honeybees but also other bee species.

Aim for your garden to have as long a flowering season as possible. Bees are mostly active between March and October, but overwintering bees – queens and some species that hibernate – may emerge in warm weather at other times of year, and will need something to eat. A densely planted plot with lots of different types of open flowers, from annuals and perennials to trees and shrubs, offers the richest possible site for all foraging insects, bees included.

▼ Different bee species tend to prefer flowers that suit their 'reach': longer-tongued bumblebees go for deeper flowers.

Plants bees like

Bees generally prefer simple, open flowers. The more complex hybrids with double layers of petals can be too complicated for them to negotiate (and don't always offer the nectar and pollen bees need). Some non-honeybee species, however, have especially long tongues and plants such as snapdragons and foxgloves, with their downward-facing, tube-shaped flowers, are their favourites.

Most bees like the fragrant herbs such as hyssop, lavender, fennel, thyme, borage, chives, mint and comfrey. Many herb flowers 'refill' with nectar especially quickly (meaning that even if a bee has just taken a quantity of nectar, the flower will already have replenished its nectar stocks when the next visitor comes along a minute or two later).

When you're buying new plants, try to buy organic. A surprising number of non-edible plants are raised using pesticides before they reach the garden centre, and you don't want any chemical residue to pass on to pollinating insects. Organic ornamentals are now becoming more widely available.

Offer a drink

Bees need water in hot weather, but often birdbaths and other sources are too deep and they run the risk of drowning while drinking. Put out a shallow dish of pebbles and fill it with water – then insects can perch on the stones to drink.

Avoid pesticides

Try to avoid pesticides altogether – many gardeners now manage to do without them completely, and this is good news for all insects. If you feel you have to use a permitted pesticide as a last resort, spray only at dusk, when the bees aren't active. Use it over as small an area as is feasible, and avoid spraying plants in flower altogether if possible, as they're the ones the pollinators will visit.

OFFERING SHELTER: MAKE A BEE HOTEL

A 'bee hotel' offers shelter and a nesting site for pollinators, including bumblebees and solitary bee species. They're available commercially (pictured below), but it's simple enough to make your own. You need short lengths of bamboo cane, roof tiles and perforated bricks or pavers, plus straw, moss and twigs for filling in the gaps. The ideal shelter should be raised off the ground – ideally by at least 60cm – on wood or bricks. Arrange the bamboo lengths end-on, alternated with bricks or tiles, fill in the gaps with straw, moss and any other natural twiggy material, and add a waterproof roof made of tiles. Then leave it alone; the less it is disturbed, the more the insects will like it.

Q What works against weeds?

ARRIVING FRESH to a neglected garden may present the new owner with a crop of well-entrenched weeds. When they present as a solid, firmly rooted surface layer, it can be hard to know where to start. What's the best way to tackle them?

Traditionally, old carpet was popular as ground cover to deprive the weeds of light to grow – but more recent thinking is that it can leach chemicals, and the foam-backed type degrades into a lot of small, messy pieces that are hard to clear. Much better are sheets of heavy plastic or woven ground-cover fabric. Layers of cardboard topped with a thick covering of straw or wood chips will make good cover.

Weight any cover materials down thoroughly so that they don't blow about. If you use plastic or ground-cover fabric, you can dig a trench along each side of the plot and let the material run over the end to the base of the trench, where it can be weighted down with bricks or stones.

If you opt for cardboard, remove any parcel tape, which won't rot down quickly, and use multiple layers, then add straw or wood chips on top. You can add to this topping with any spare lawn clippings or other green material. The top layer will rot down, and will enrich and improve the soil as it smothers the weeds.

A

The good news is that a healthy crop of weeds is likely to indicate that the soil you've inherited is fertile. Meticulously digging them out is the hard-work solution, but often the simplest answer is to cover the whole area, depriving the weeds of light – over time (it will take about a year), all but the hardiest will give up.

How can I get out of weeding my garden?

It's a common occurrence when gardening: you start weeding and an hour later, congratulating yourself on a job well done, you stand up – and promptly spot a patch of weeds you didn't get to. It's best not to be too neurotic about removing every single weed: as a gardener, a philosophical approach will stand you in better stead.

All gardens need some weeding, and you'll never get rid of weeds entirely: they can be managed, but not eradicated. They're opportunists, they move in when they see a space. Aim to minimise weeds, and plant your garden so as to leave them little room to take over.

If you close-plant beds, leaving them without any bare soil, you will largely squeeze weeds out, and the few left will be disguised by the crowd of other plants. This approach does require some monitoring, though – some weeds, for example bindweed, are happy growing through other plants and have to be pulled out before they go too far; others – chickweed, for example – will sneakily grow under the foliage cover and you need to hunt them out.

Plant your beds so that they overhang their borders: weeds love empty edges, so fill your beds to their limits to avoid them getting any space. Leave just enough room under the overhanging foliage to run your hoe through, quickly, easily and often.

GOOD WEEDING HABITS

When you're out in the garden, even if you're just enjoying a cup of coffee for 10 minutes, get into the habit of picking out a few weeds while you're there. This will help you stay on top of the job. Always, always pull off the heads of any weeds that you see have got as far as flowering, even if you don't have time to dig or pull the roots out – it means that they won't have the chance to set seed and multiply.

Q Does anything grow under trees?

IT'S A COMMON SITUATION: the garden is growing beautifully everywhere except in the area under the tree or trees that make a focal point. It's not the friendliest location for plants, so what can you grow there to fill in the gap?

It's easier to find plants that will do well under deciduous trees; evergreens (such as holly) or conifers (such as yew) are more challenging, as they are shady year-round. Such plants need to overcome the fact that the ground tends to be dry, with poor soil – because the tree has taken the nutrients – and may have roots near the surface, making it hard for gardeners to dig to any depth.

If the soil is bare under the tree, start by mulching it, using a layer of wood chip, bark or leaf mould. Even if you can't dig extra organic matter in, mulching will help to enrich the soil. Lay a layer of mulch around 5–10cm deep, but don't mulch right up against the trunk – this can encourage rot. Leave a space of at least 20cm between the mulch and the tree.

Plants that thrive under deciduous trees benefit from some light and rain in late autumn and winter when the tree has shed its leaves, and often do most of their growing in the early spring, so that they are already well advanced by the time the leaves of the tree come out, casting them into shade.

If you want to grow plants under a conifer or evergreen, give them as good a start as possible by putting them into holes that are as large as you can excavate (tree roots allowing) and surrounding them with additional compost rich with organic material. Then water generously and check on them regularly.

Grape hyacinth, *Muscari*

A

The ground under trees is difficult as it is usually both dry and – depending on whether the tree is deciduous or not – shady for part or all of the year. But there are various plants that will work in this situation, and it is also usually possible to improve the soil.

A FEW PLANTS THAT WILL WORK IN DRY SHADE

Bulbs and corms

- Snowdrops, *Galanthus*
- Grape hyacinths, *Muscari*
- Daffodils and narcissi, *Narcissus*
- Autumn-flowering cyclamen, *Cyclamen hederifolium*

Any or all of these will flower successfully under both deciduous and evergreen trees provided they've been planted in enriched soil. In very dry conditions, they may need additional watering.

Perennials

- Lady's mantle, *Alchemilla mollis* – light-green leaves with scalloped edges and sprays of small yellow flowers carried above them.
- Christmas rose, *Helleborus niger* – early flowers, with varieties that are white or deep mauvy-pink, or hybrids that range from red, green or yellow, all the way to very dark purple and almost black. The originals have simple flowers, but many hybrids are fancier, with deep-coloured spots or double flowers. Handsome deeply cut, dark-green foliage lasts after the flowers are over.
- Epimediums, *Epimedium* – there is a range of these available, most with pretty heart-shaped leaves that are often bright green or edged with a red flush. The flower stems are carried above the foliage, bearing small, pointed flowers in a range of colours from white to yellow and pink. They are very resilient, though they prefer acid soil.

Ferns

- Male fern, *Dryopteris filix-mas* – good-looking upright fronds that grow in an inverted cone shape. The leaves take on a bronzey colour in autumn, then die back in winter before new fronds uncurl from the fern's 'crown' in spring.
- Hart's tongue fern, *Asplenium scolopendrium* – wavy leaves that are rich green with a distinctive 'tongue' shape. They tolerate very poor (or almost non-existent) soil, often growing out of dry walls.

Hart's tongue fern, *Asplenium scolopendrium*

How much should I mulch?

MULCH IS USED AS A TOP LAYER on soil to help it retain moisture in hot weather, to protect plant roots from the extremes of hot and cold and to quell weeds. Mulches also give a neat finish around plants and are typically used under trees, shrubs, fruit bushes and climbers. But what should you use, and how much of it?

Biodegradable mulches, such as leaf mould, compost and wood chippings, really help the soil. Although it would be hard to overdo it, generally you want to make sure that you aren't heaping mulch too high around smaller plants.

Leaf mould, compost and wood chippings are biodegradable mulches; so, too, are rotted-down manure, seaweed, straw and used mushroom compost. All offer at least some nutritional benefits for the soil.

Non-biodegradable mulches are also available, including gravel, pebbles and slate and stone chippings. These will give surface protection to plants and a neat finish to the planted area, and will also discourage weeds, but they won't nourish the soil underneath.

Sheet mulch

Finally, there are so-called sheet mulches: these come as flat sheets of 'membrane', usually black and porous. Make sure only to use the biodegradable type; plastic or synthetic mulches – particularly old carpet or PVC sheet – degrade quickly into thousands of minute pieces that are impossible to remove and that pollute the soil.

Sheet mulches are generally laid on new beds, sometimes for a year or two before planting, to kill any weeds, or directly before planting, to keep the weeds down while new plants, planted

Prunings from shrubs and trees can be chipped or shredded, making a long-lasting woody mulch which rots relatively slowly.

through slits in the membrane, get established without competition. They can keep an area weed-free for some years but do have a utilitarian look, so tend to be used most often on vegetable and fruit-growing plots.

How and when to mulch

It's hard to over-mulch, but remember that the benefits should be directed to the soil surrounding plants and it should not be heaped up too high around smaller plants or the stems or trunks of shrubs and trees. Most gardeners use biodegradable mulches in 5–10cm deep layers.

Mulch is usually best applied from midwinter onwards, before weeds start to grow again. Winter's end is usually the ideal time to apply mulch to soil. Extra nutrients are unnecessary from the end of summer to late winter, when most plants are dormant.

Watering and feeding

Gardeners need to water for longer in mulched areas to ensure the water gets through the mulch and into the soil, although mulches are generally used for trees and plants that are longer-lived and that may not need much or any watering once estabished. Feeds are also usually applied directly through mulches, although solid fertilisers, in the form of pellets or powders, may be best added in late winter, shortly before mulching, to give the last of the winter rain the chance to wash them into the soil.

SEASIDE BENEFITS

As a raw material, seaweed has been pressed into service as mulch by seaside gardeners for centuries. It's popular, too, with gardeners who prefer to avoid both commercially produced fertilisers and animal products. Most landowners and local councils won't object to you collecting a few sacks direct from the beach, provided that you ask first. Although it's traditionally simply left as a thick layer of mulch on the soil surface, it can get quite smelly as it dries out, and modern gardeners usually dig it in as a soil conditioner – it contains potassium and magnesium, and plenty of trace elements, and it can be used just as it is, without leaving it to rot down before applying.

What's killing my new plants?

A classic case of pot-bound roots; don't buy a plant if you can see roots sticking out of the drainage holes of its pot.

IF YOU'VE ONLY RECENTLY planted something special in your garden, it's upsetting when it fails to thrive, despite the care you've lavished on it. And it's worse when it suddenly dies on you. How can you find out what went wrong and ensure that it doesn't happen again?

It may sound like an obvious question, but are you sure it's dead? Even when a plant's leaves have wilted, browned or dropped, it's not necessarily a goner. Try snapping a small stem from it – if it breaks easily and looks brittle and dry inside, the signs aren't good; if it bends, breaks reluctantly, and looks moist and greenish inside, the plant is still alive. Even if the stem is brittle and brown, it's worth going down to the roots and seeing how they look: if they are strong and flexible, it still has a good chance of survival. Darkened, brittle or soggy-looking roots, however, mean that it may be time to give up on it.

When the plant was planted recently, ask yourself the following questions:

If it came in a pot, did you loosen the root ball and tease out the roots slightly before planting?

Did you check which soil this particular plant likes? Is it an acid-lover accidentally sited in a limy soil?

Was it over- or under-watered?

If it came straight from a nursery, where it was kept in sheltered

Even when a plant's foliage has died, it's always worth checking the roots for signs of life – it's possible that it could be revived.

conditions, did you harden it off before planting it out?

Has there been a recent sudden drop in temperature, or even a frost?

Was the plant given a little time to acclimatise to its new conditions, in a sheltered spot or a cold frame?

It's not unusual for one or more of the above circumstances to kill a plant, which is always at its most vulnerable when it has just been planted in a new site. Gardeners sometimes suspect a pest or disease when the cause is more commonplace. That great garden villain honey fungus, for example, will rarely strike a newly sited plant – it takes time to take effect.

Use the guarantee

Many nurseries guarantee their plants against sudden death shortly after purchase. This is because some not-so-obvious problems, including water moulds such as phytophthoras, are very hard to eradicate in nurseries and may cause problems in their stock. If you can't explain what went wrong, take the plant back to the nursery and see if they will give you a replacement. If they do, check the plant's needs carefully with them, and try again.

Don't beat yourself up

Remember that every gardener sometimes has a plant fail; they learn from it (perhaps they put a sun-lover in deep shade, or didn't harden off a plant before putting it out) and don't make the same mistake again.

A There are many possible causes for a sudden plant death. You can do a post-mortem to see which might have caused your fatality and which can be ruled out. Bear in mind, depending on what you find out, that it may not be wise to plant a replacement in exactly the same site.

FROM THE ROOTS UP

If the stem test failed but the roots of the plant still look healthy, you may be able to get it to regrow from the roots. Dig the plant up, then put it in a pot of moist, not wet, compost. Make sure you ensure that there is enough room for the roots. Then trim back the stems until you come to living wood – this may be a third or two thirds of the way down the stem, or very close to the base. Put the pot in a sheltered place – ideally in a cold frame – and keep an eye on it. If it begins to show signs of regrowth, wait until it's got a good start before planting it out again.

Q Why don't worms like my soil?

WE'RE SO USED to associating healthy soil with the presence of worms – and so used to hearing how important they are – that it can be disconcerting when you dig over a few square metres of soil without finding any. Why aren't they there, and can you do anything to encourage them into your garden?

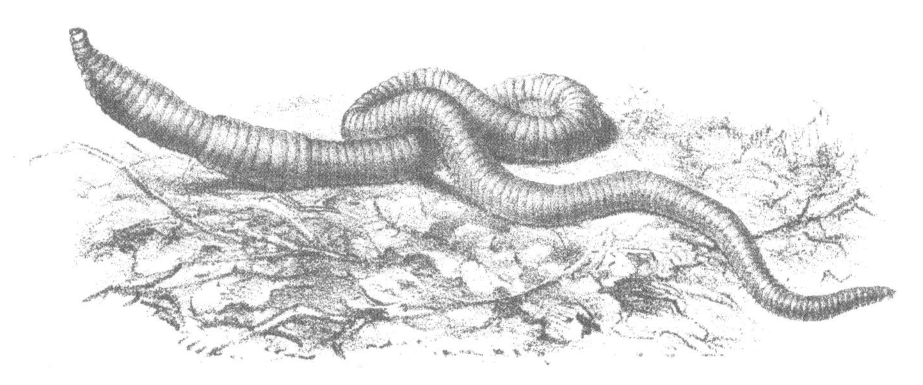

A Worms like moist, nutrient-rich soil full of organic matter. They don't like very acid, very sandy or very wet soils. However, just because you didn't see any in your soil, this doesn't necessarily mean that they aren't there.

Worms are important because of the job they do in refining and aerating the soil; some species are active up to 2m below the surface – invaluable whatever sort of garden you have, but especially useful if you operate a no-dig policy. And most soils do have worms: a study at Rothamsted Research Centre in Hampshire found that even poor soil can support as many as a quarter of a million earthworms in an acre, while a rich soil might have seven times as many. In dry, hot weather, worms will go much deeper underground, reappearing after the first day or two of rain showers.

Bringing them to the surface

If you don't spot worms when you're digging in hot or dry weather, don't immediately conclude that your garden doesn't have any – they need moisture to breathe through their skins, so wait until the day after a heavy shower of rain, then dig again: the chances are that this time you will find they are there after all, just below the surface.

If there still aren't any worms, it may be that your garden needs a feed. Dig in as much compost, well-rotted manure or leaf mould as you can, to make a more worm-friendly habitat.

FLATWORM VERSUS EARTHWORM

In the late 1960s, the New Zealand flatworm arrived in UK gardens – it probably hitched a lift in imported pot plants – and was soon found to feed on native earthworms. As its name implies, it's flat and brown with pointed ends, up to 20cm long when it's moving (although it's usually seen coiled up, looking much smaller) and very slimy. It was followed by Australian and Brazilian flatworms (both of which are orange, rather than brown), but although none of these incomers is good news for gardeners or native earthworms, their impact on UK gardens and their earthworm populations has not been as bad as was originally feared. Very damp habitats suit flatworms better than dry ones, and they have had a more serious effect on Scottish gardens than on those south of the border, where flatworm and earthworm populations seem to have reached a balance after an initial drop in earthworm numbers. Researchers have also found that flatworms have some predators of their own: lawn beetle larvae eat them, and it's believed that centipedes, too, may have them on the menu.

Common earthworm, *Lumbricus terrestris*

If you find one of these unwelcome worms in your garden, squash it. And do be cautious about giving plants to other gardeners – you don't want to spread the population.

New Zealand flatworm, *Arthurdendyus triangulatus*

Is there anything I shouldn't add to my compost heap?

SOME PEOPLE TREAT THEIR COMPOST like a precious recipe that 'cooks' only when put together with very specific ingredients; others seem to put everything but the kitchen sink onto the heap – and still produce good compost. How crucial are the do's and don'ts of home-made compost?

▲ Plants that have leathery or waxy evergreen leaves, such as azaleas, *Rhododendron*, will take ages to rot down unless they are shredded first.

The main 'don't' for compost is adding non-vegetable kitchen scraps – meat, fish, bread and so on. This is definitely to be avoided: you run the risk of attracting rats. Too much citrus is also often cited as a no-no, although it will rot down in the end. It may be very slow to do so in a small compost heap where additions are little and often. The same goes for very thick evergreen leaves, such as bay or rhododendron: they will rot, but they can take a very long time.

Keeping the balance right

The most important thing for compost is to get a good balance between the different materials – the best mixture is composed of fairly even amounts of 'green', nitrogen-rich material (grass clippings, leaf trimmings, weeds) and 'brown', carbon-rich material (woody stems, dead leaves, cardboard, wood chippings, straw pet bedding and so on). Too much green waste, in particular a lot of grass clippings, can result in a wet and slimy mix; conversely, if there's an imbalance towards the 'brown' side, the heap can be terribly slow to rot down.

▼ Dead leaves are fine in compost, although if you have a lot of them, you may opt to make a separate quantity of leaf mould (see p144).

DEALING WITH ROOTS AND WEEDS

Seedheads and the roots of tough weeds or diseased plants shouldn't be added as they are, as the temperature of a domestic-sized compost heap is unlikely to get high enough to eliminate their diseases or break the plants down. Instead, put them in a bucket of water and let them rot down for a week or two before adding the mixture to the compost.

Weeds may resurface later if you add them to compost; rot them in water first, so their roots and seeds don't survive.

How long does it take?

Most material takes between six months and two years to compost thoroughly. Regular turning (mixing up the 'layers' of the heap with a fork) is very helpful, but the compost will still rot down eventually even if you don't. If some of the heap is mature compost while some is still relatively raw and not yet rotted down, take out and use the compost that is ready and use the opportunity to give the rest a thorough mix.

A You should definitely avoid adding a few things to the compost heap, such as meat and fish, but if you follow a few simple rules, the composting process will deal with quite a lot of other waste.

Eggshells, banana peels and coffee grounds can be composted, although they may take two years or more to rot down.

Q Can I make quick compost?

YOU MAY HAVE HEARD that it takes two years to make compost. Is there any way that you can help to speed it up?

There's no mistaking the rich, earthy smell of compost that is ready to use. It should also have a deep brown colour and a crumbly texture. Don't worry, though, if it still has some twiggy bits in it – this is normal, even in mature compost.

Let the worms do the work

If your garden is really tiny, with no space for a compost heap or even a compost bin, a wormery is one way to deal with smaller amounts of both kitchen and garden waste while also creating a useful tonic for the garden. The worm compost produced makes a good soil conditioner, while the liquid drained off can be diluted and used as a fertiliser. All a wormery needs is a sheltered corner in the garden, and a little day-to-day care.

Wormeries are inexpensive plastic boxes, around 60cm high. They usually have two compartments, a top one in which the worms live and work (which is where you put your scraps or garden trimmings), and a lower one that collects the liquid that drains off. The more elaborate versions also contain stacking trays which make it easier to remove the compost as it is ready.

THREE WAYS TO SPEED IT UP

- Enclose your compost with 'walls' – open compost heaps take longer.
- Shred or cut up material such as woody stems and plant waste into smaller pieces before adding them: they will rot down faster than if you leave them whole.
- Turn compost thoroughly and regularly, mixing it all the way through from the base of the heap. This introduces air, which is key to the whole composting cycle. If you notice as you turn that it is dry, add some water; if wet, an addition of straw or scrunched-up sheets of newspaper will help.

> There are various ways to hurry compost along, but only to a certain degree: even at its fastest, it will take around six months to be usable.

Composting worms

The wormery's inhabitants aren't earthworms, but composting worms. They are smaller and redder than earthworms, include several different species and can also be known as brandling or tiger worms. They're put onto a layer of material – old compost is fine for this, although the manufacturer usually supplies a layer of coir with a wormery, and the worms' food is placed on top. This includes vegetable trimmings from the kitchen, garden weeds and so on.

The amounts need to be fairly modest and the smaller the material is chopped up, the easier it is for the worms to deal with. Most of the material that would be suitable for a compost heap can also be put into a wormery, with the usual caveats about citrus – the worms don't like too acid an environment – and too-large pieces of woody or twiggy material.

Check every so often to make sure that the top layer of material is being dealt with. The wormery is ready for more material when you see a few worms on top of the most recent 'feed'. Usually a wormery needs feeding once or twice a week.

The temperature is important – worms do best at 18-26°C; in lower or higher tempeatures they will become less active. Worms need to stay moist, too; this isn't usually a problem if they're fed regularly, as organic matter offers plenty of moisture.

If the wormery doesn't have a tray system, you empty it when it is full, separating the worms from the compost and replacing them in the base of the top compartment (they tend to group just under the top layer of material, so this isn't as tricky as it sounds) – then simply repeat the cycle.

▼ A wormery with several tiers or trays allows you to remove small quantities of compost once it's ready, as well as draining off liquid manure.

Q How can I stop birds eating my grass seed?

GROWING A LAWN FROM SEED is cheaper than turfing and, once it gets started, can be easier to maintain, too. But pigeons and other birds are also big fans of grass seed, so how can you ensure the seed gets the chance to grow?

A If the seeded area of your garden is quite small, you can cover it with well-fastened netting or fleece, so that the birds can't get at it. For a large area, where this wouldn't be practical, your best bet is to try to scare them away instead.

Sown in mild, moist weather, grass seed should need only three to four weeks to become established, after which the depredations of birds won't be a problem.

Prepare your ground

This is harder work than actually sowing the seed – in dry weather, weed, dig and fork the surface, ideally to a depth of 20cm, until you have a level, smooth surface. When you've done this, walk across it a few times – in a pigeon shuffle so you cover maximum ground without trampling it – leaving it loose on top, firm below. Finally, leave it for at least two days before you sow. It's well worth doing the preparation thoroughly: you'll end up with a nice level lawn that's free of perennial weeds.

PIGEON SCARING

Keep birds right away from your future lawn by setting up scarers, such as old CDs that shimmer and reflect. You can hang these from stakes, or string them up on lines fastened to stakes at each corner of the sown area, surrounding it completely. You can also buy strips of reflective anti-bird tape that both flashes in the light and makes a high humming noise. This should keep the birds away.

Can anything stop ants?

ANTS' NESTS ARE OFTEN PRESENT in the garden, even if you don't see them. A heap of fine, freshly excavated earth on the lawn may be the only sign of an ants' nest underground, and they will occasionally make a home in a large container, too.

Ants' nests are usually best left undisturbed – loose soil on lawns or the edges of beds can simply be swept away. A nest in a large planted container may disrupt the plants' roots and thus do some damage, but is often a sign of under-watering. Setting up a more regular watering regime will encourage the ants to find another home.

Ants are universal in gardens and they don't cause much harm except some slight and occasional damage to plant roots. Their minor activities shouldn't damn them with the name of 'pest', so they don't really need to be stopped; it's better to live alongside them.

DO ANTS ENCOURAGE APHIDS?

Ants drink the honeydew – the sweet liquid excreted by aphids – as a food, and will stroke greenfly with their antennae to encourage them to produce it: that's why it's sometimes said that ants 'milk' aphids. Ants may also see off aphid predators, such as ladybird larvae, to protect their asset. Most gardeners have their own ways of dealing with aphids, however, and the presence of ants doesn't prevent them from spraying greenfly colonies with soap, or squashing them by hand.

▲ Ants benefit from the presence of aphids, but they they don't encourage them to your garden; the aphids arrive under their own steam.

Q Why does my new lawn curl up at the edges?

THE DEFINITE PLUS OF A LAWN LAID WITH TURF is that you get virtually instant results. You need to follow a few rules to make sure the lawn lies flat, though: otherwise, you may find that it curls up like a dried-out sandwich once it's in place.

The preparation for laying a turf lawn is exactly the same as for sowing a lawn – with the obvious advantage that, once laid, a turf lawn is ready to walk on within a few days. An additional step for a turf lawn is to rake in fertiliser after your soil has been prepared and before you lay the turfs down.

The main reason for an October-to-March window for laying down turf is that it needs to be kept moist, so it's best not to put it down in summer when it's more likely to dry out. Laying it in autumn or winter means that the lawn gets plenty of chance to root into the underlying soil before mowing time comes around in later spring.

A You may have bought your lawn at the wrong time of year. Make sure turf is laid down at some point in the cooler months. Prompt action is also vital to prevent future curling: ideally the turf will go down the day it arrives, so some preparation is needed.

LAYING THE LAWN

- Lay the turfs working from one corner of the lawn area, standing on boards to avoid dents in the freshly laid grass, and laying outwards onto the prepared soil. Lay them like bricks rather than a grid – that is, staggering the edges.
- Place any smaller pieces of turf near the middle; if you lay them at the edges, they are more likely to dry out, shrink and then curl.
- Have a bucket of fine, crumbly soil at hand so that you can fill in any small hollows or gaps between the turfs that become apparent as you lay the grass down.
- Rainfall will usually keep the new lawn moist. Water if there's a prolonged dry spell (say, for longer than 10 days).

Will nematodes get rid of pests?

NEMATODES ARE THE MODERN gardener's alternative to many old-style pesticides. They're all members of the Nematoda, a large family of worm-like organisms, abundant in a wide variety of environments, from the tops of mountains to the depths of the sea. Nematodes are also present in the soil in your garden. It is some of the parasitic members of this prolific family that can be put to good use.

Nematodes can work very well if used at the right time in the right way on the right pest. They are usually most successful on pests found in the soil, such as slugs; they tend to be less effective above ground on prey such as caterpillars, because they don't survive in dry conditions.

To be effective, nematodes need moist, warmish conditions, so they are best used in late spring or summer. Although the most popular nematodes tend to be those used against slugs, most beneficial versions are applied in the same way. Mix the contents of a packet (in which the nematodes are far too small to be visible) with tepid water and pour the mixture on to the affected area of the garden. Once they are in the soil, they will home in on their natural host. Different nematodes have proved very effective on a range of pests, for example, vine weevils, chafer grubs and carrot fly, as well as on slugs.

Because they are living creatures, nematodes are usually bought by mail order – they can't live indefinitely in packets on a plant nursery shelf.

NOT ALWAYS THE GOOD GUYS

Nematodes can be pests themselves: plenty are parasites that damage plants by piercing their cells and sucking them dry, whereupon the plant may respond by developing lumps, bumps and lesions. Rest assured, though, that those sold for garden use are completely harmless to plants, pets and people.

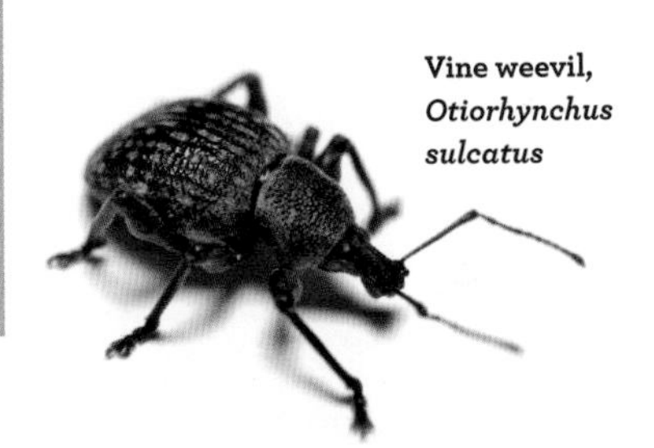

Vine weevil, *Otiorhynchus sulcatus*

Can I ever use chemicals?

BACK IN THE 1950S AND 1960S, most gardeners were happy to turn to chemicals to solve most common garden problems: pesticides, fungicides and weedkillers were in almost every garden shed. Today, environmental and safety legislation is much more stringent in many countries, and although those pesticides that are legal and available are effective and as safe as they can be, many gardeners still prefer to manage without them.

There are alternative solutions for many pests and diseases, so do your research thoroughly before thinking about going down the chemical route.

Pesticides should always be a last resort – other methods of controlling pests, diseases and weeds should be tried first.

Options without chemicals

Check plants frequently. If a pest or disease is spotted early enough, leaf-picking or insect-squashing may be all that is necessary.

Think about growing favourites in a different way – for example, if the only way you can have hostas in open soil is with the lavish use of slug pellets, grow them in pots instead, where they can be defended with suitable barriers such as copper tape or wool pellets.

▼ Hostas, which are particularly popular with slugs, are easier to protect without the use of chemicals if they are grown in pots.

Many types of plants are now bred to resist some diseases – for example, apple trees with strong resistance to scab diseases are now available, as are groundcover roses which don't tend to get blackspot.

Look at all the alternatives to chemicals – biological controls, traps and barriers. The options available depend on the problem: nematodes are often strongly effective for soil-based pests, for example, while sticky traps work well for slugs in greenhouses, and pheromone traps for a whole range of other pests. Fleece will act as a barrier for mini-pests such as flea beetles, while insect-proof mesh in the vegetable garden will guard against leek moth and carrot fly.

TIPS FOR USING CHEMICAL TREATMENTS

- Identify the problem accurately and check that any pesticide is appropriate for the specific problem before buying. Many nurseries can offer good pesticide advice, so take advantage of it.
- Aim to use pesticides allowed for organic gardens, such as soap-based insecticides. They often give sufficient control, particularly if pests have been spotted early.
- If you have to spray, avoid hot, dry weather and do it at dusk when most insects are not on the wing. Adjust the sprayer to a larger droplet that won't drift, and don't spray when there is a breeze.
- The greatest environmental risk is actually posed by the disposal of pesticides, so buy only as much as you need, and take any unwanted pesticides to an official disposal site.

▶ If you're spraying chemicals, always use a spray attachment with a variable nozzle and set it to the 'coarsest' (largest-droplet) setting.

Q Should I worry about woodlice?

GARDENS ARE USUALLY HOME TO WOODLICE, and plenty of them. Sometimes when you pick up a flowerpot, what seem like hundreds scurry around in circles before running off. But exactly what are they, and should they be regarded as garden pests?

A

No, woodlice aren't significant pests in the garden. However, you do often find them around the sites of damage caused by other pests.

Because woodlice live off decaying organic matter, they sometimes attract suspicion when they're found clustered around the site of damaged plant roots or vegetables. But this is because they're opportunists – they don't initiate the damage themselves.

LOUSE LIFE

You might suppose that woodlice are insects, but they're actually crustaceans – the land relatives of crabs or lobsters. Most gardens offer plenty of the rotting plant matter they live on, and they may be especially abundant in the compost heap. They're nocturnal: during the day they rest up somewhere dark and damp, but at night they come out to feed. The female lays eggs but keeps them in her body until they hatch, when the tiny new woodlice are moved to a brood pouch for a few days before they leave their mother and disperse. In spring you can often spot the minute young, pale-grey replicas of their parents, clustering with the adults under plant pots or bricks.

A mixture of adult (darker grey) and juvenile (pinkish-grey) woodlice, colonising rotten wood, one of their most popular habitats.

Q Are spiders garden friends or foes?

IF YOU'RE ONE OF THOSE people with a visceral distaste for spiders (the scuttling motion! The many legs!) then you may feel that, whatever their role in the garden, you don't want them around. But what part do they actually play in garden life?

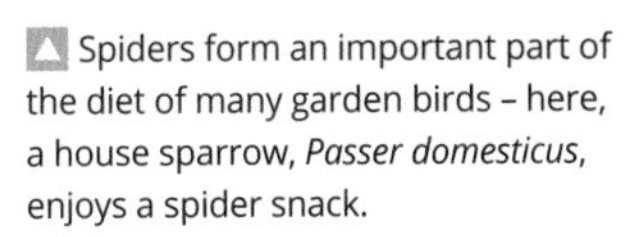

Spiders form an important part of the diet of many garden birds – here, a house sparrow, *Passer domesticus*, enjoys a spider snack.

A Across the board, spiders do more good than harm in the garden. They are an important part of the food chain, eaten by birds and small mammals, as well as eating many insect pests themselves.

Of course 'good' insects, such as moths or hoverflies, may be lost to a spider's web, but against that may be counted the many others that aren't friends to the gardener, including aphids, thrips and more. And while you're used to seeing spiders up in their dew-drenched webs in autumn, waiting for their victims to stop by, there are many garden spiders that you won't ever notice, because they're not web-spinners but hunters or ambushers, lying in wait for their preferred prey and then leaping on it or chasing it down.

OUTNUMBERED

Whether you like them or not, there are an awful lot more spiders than there are of us: some studies suggest the spider-to-human ratio in the UK is about 500,000 to 1, with more than 650 native species. Even this total is easily outdone by the USA, which boasts over 38,000 species.

A wolf spider, one of more than 2,300 members of the Lycosidae family found all over the world. The name reflects their keen hunting ability.

Q How can I bring my soil back to life?

THE HOLY GRAIL OF MOST GARDENERS is rich, crumbly, dark-brown soil, moist without being wet, worm-rich and evidently just waiting to get on with growing things for you. But what if you've inherited a plot with less-than-ideal soil – what changes can you make in it, and how?

Healthy earth is teeming with micro-organisms far too small to see: just a gram of soil may be housing as many as a million of them. Between 3 and 6 percent of healthy, productive garden soil consists of organic material, but overused and underfed soil may have just 2 percent or less of this vital component. The simple addition of organic material, whether compost, manure or leaf mould, immediately introduces new micro-life to tired soil.

▼ If you can make enough of it, home-made compost is ideal for reviving weary soil; if not, you can alter it with one of the alternatives.

A Organic material can make every soil better, no matter how poor it may be to start with. And if you're prepared to put some work in, you can hugely improve your soil.

What to use

If you don't have enough home-made compost to feed your garden, there are plenty of inexpensive alternatives to try. Spent mushroom compost, manure from a local stable (though make sure it's well rotted down before you use it) or compost from an outside source – local councils often

sell it at refuse and recycling centres – are all possibilities.

If you can't – or don't want to – dig the material into the soil, any of these can be applied to the garden as a mulch. The nutrients may act more slowly than if the organic matter were thoroughly dug in, but they will gradually leach into the soil below.

PIONEER OR COVER CROPS

If your soil is very poor, if it consists of dense clay, or is on a site where it has been heavily compacted, consider a traditional remedy: plant pioneer crops, also known as cover crops or green manure. With this method, you plant a crop for the sole purpose of improving the soil: nothing is as effective as plant roots in aerating and breaking up compacted soil, and a season of growing a tough, resilient plant or mixture of plants can have an almost miraculous effect in bringing soil back to life. Some possibilities are mustard, which grows both fast and deep; fodder radish (not the same as the table radishes we eat) – which is extremely deep-rooting; buckwheat, for dry, sandy soil, or ryegrass for an overwintering option. When the crop has done its work, you can dig it in or cut it before it sets seed and leave it on the surface as a mulch.

White mustard,
Brassica alba

Buckwheat,
Fagopyrum esculentum

What type of soil do I have?

SOIL VARIES HUGELY across different landscapes, and what type of soil you have in your garden will impact how easy it is to work with, how much organic material you'll need to add and, in a few cases, what you can grow. If a handful of your soil feels gritty, does that mean it's sandy? And, if so, what are the repercussions in gardening terms?

Experienced gardeners can usually tell you your soil type from looking at it and from the way that it feels in the hand. For instance, a gritty handful means that it's sandy, which also means that it's quick draining, light to work with, but may need extra feeding to bump up its fertility.

Soil is traditionally divided into six different types; sandy is just one of them. Here are the 'pure' categories with their key characteristics, although garden soil will rarely fit neatly into a single category – it is much commoner to find that you have a mixture, such as a silty-clay loam, or a sandy loam.

Clay

Clay is 'heavy' soil, made up of very small particles that make it dense and slow to drain. Nutrients don't easily wash out of it, but it needs breaking up and aerating. In spring it is slow to warm up, but it can bake dry in a hot summer. Clay soil feels smooth, clumps easily and is slick and sticky when wet.

Sand

Sand is 'light' soil and has the largest particles of any soil type, so it drains quickly and nutrients may wash out of it. It's light and easy to work with and warms up quickly in spring. Sandy soil feels gritty and even when it's damp, won't clump together easily.

You can buy hydrangeas that will manage in a range of soils, although fertile, free-draining types suit them best.

Silt

Made up of mid-sized particles (smaller than sand, larger than clay), silty soil retains both moisture and nutrients better than sand and drains quite well – but it is easily compacted. Silty soil feels slightly slippery to the touch, and even when damp, doesn't clump easily.

Loam

The gardener's ideal soil type, loam is a combination soil, containing a fairly even balance of silt, sand and clay, along with a good proportion of organic material. It's both fertile and easy to work. Loam is crumbly in the hand, and clumps easily but loosely when damp.

▲ These profiles from the Trials Field at the RHS garden, Wisley, Surrey, show the variety between different soil types. They are: (A) loam-rich soil, (B) new-build soil, (C) sandy soil and (D) soil taken from an orchard.

Peat

Although peat always appears on soil lists, it's actually very rare to find it in gardens, and wild peat habitats are generally in decline (which is why many gardeners prefer to buy peat-free compost). Peat soils are extremely rich in organic matter. They are moisture-retentive but tend to be low in nutrients and are ideal for growing ericaceous, acid-loving plants. Peaty soils feel slightly spongy if squeezed.

Heather,
Erica

Chalk

Chalky soil can be either light or heavy (the latter has a higher proportion of clay), but is always highly alkaline, which means that it won't support ericaceous plants, such as most forms of heather (above right).

Clay, sand, and chalk can all be improved by adding organic material – silt and loam soils, which tend to be naturally fertile, will benefit from this, but don't need it to the same degree. Should you, very unusually, have peat soil, it's almost the only type that doesn't need extra organic matter, although it will benefit from fertiliser.

Why are my leaves powdery?

A PREVIOUSLY HEALTHY ROSE has developed a soft, white coating on its leaves. They're also distorted and shrivelling up at the edges: they look sick. What's wrong, and how can it be fixed?

This sounds like a classic case of powdery mildew, which is a fungus infection. Both powdery and downy mildew can be a problem in gardens, but, confusingly, despite the shared name, the two types aren't related.

With withered patches on top and mould underneath, these rose leaves show classic symptoms of downy mildew.

What do mildews affect?

Each of the two terms, powdery mildew and downy mildew, refers not to a single infection, but to quite a wide group. Each infection tends to affect only a very specific group of plants – the powdery mildew on the rose, for example, will not be the same one as might attack an apple tree, and the one that affects the apple tree will not be the same as the one to be found on a climbing honeysuckle.

What's the difference between powdery and downy mildews?

The two groups have quite different symptoms – powdery mildew results in a white-dusted, distorted appearance, while downy mildew tends to create dry, withered-looking patches on the top of the plant's leaves and a mouldy substance on the undersides. Both types are fungus infections, spread

Spots of white fungus on the top surfaces of rose leaves that are beginning to wrinkle and wither signal the presence of powdery mildew.

by spores, so it's especially important that the foliage of any affected plants is burned rather than put into the compost heap.

How can mildews be treated?

In the case of powdery mildew, there are a few chemical sprays specific to a plant or a group of plants, but many gardeners prefer to give the plant extra TLC in the hope of strengthening it enough to beat the infection. If the problem is spotted very early on, cutting out the infected leaves and shoots immediately may work.

Generous mulching and very regular watering helps, and a home-made spray can also be effective (see box).

There is currently no chemical treatment available for downy mildew; prevention is the best cure. It's attracted by damp or high levels of humidity, so avoid watering plants in the evening and make sure that plants kept in greenhouses or under cover are as well ventilated as possible. The spores of downy mildews can spread in the soil as well as in the air, so if it has attacked any of your plants, avoid repeat-planting in the same spot – fill the space with something different.

Some plant cultivars resistant to mildews are on the market, so check what is available when you're shopping for plants in a specific group.

A HOME-MADE ANTI-MILDEW SPRAY

All kinds of home cures have been devised to fight powdery mildew, using raw materials ranging from milk to garlic, but those containing either sodium bicarbonate – plain baking soda – or potassium bicarbonate seem to be the most reliably effective.

- Add 1 tablespoon of sodium bicarbonate or potassium bicarbonate to 4.5 litres of water and stir until dissolved.
- Add a teaspoon of liquid soap and two tablespoons of vegetable oil to the mixture and stir thoroughly.
- Decant the mixture into a hand sprayer, shake well, and it's ready to use.
- Spray the plant regularly, repeating after rain and spraying over new growth as it emerges.

Regular store-cupboard ingredients, such as sodium bicarbonate, can make an effective treatment against mildew.

What is leaf mould for?

ALTHOUGH IT'S OFTEN CONFUSED WITH COMPOST by the inexperienced, leaf mould is an altogether simpler matter. To make it, leaves are swept up and kept until they have thoroughly rotted down. But what's the best use for the result?

It would be easier to ask what leaf mould isn't for. It's one of the best general soil improvers available. Leaf mould can be used to sow seeds into or mixed with soil in containers, and it makes excellent mulch. The only problem with it is that most gardeners can never make or source quite enough of it.

All you have to do if you want to make leaf mould is sweep up your leaves and keep them – in larger gardens where there are plenty of leaves when autumn comes, it's worth making a leaf store. This is a simple open frame made of wood or panels of chicken wire, with staked corners. After the leaves have been piled into it, it's a waiting game: they will take a couple of years to rot down.

Even if you have a small garden with only a tree or two and without space for a leaf store, it's still extremely easy to make a small quantity of leaf mould provided that the leaves you have access to come from deciduous trees (evergreen ones may take a very long time to rot down – see box). Collect the leaves together, get a couple of sturdy plastic sacks and fill a watering can. Fill the sacks with the leaves, lightly watering the contents when the sacks are a third full and then again when they are two-thirds full. Then fill them up and tie the tops. Use a garden fork to poke a few holes in the sacks and put them away in a corner. Check after a year, and if the leaf mould isn't completely rotted, re-tie the sacks and leave for another few months. When you next look, the sacks' contents will have transformed into leaf mould, ready for use.

Leaf mould is one of the simplest – and most valuable – things a gardener can make. All you need is access to plenty of deciduous trees.

WHAT KIND OF LEAF?

Any and every leaf will make leaf mould, but how long they take to rot down satisfactorily depends on what types of leaves you have. If you're collecting from a mix of deciduous and evergreen trees, you may end up with two or more different stores. Leaf mould is so valuable in the garden that they will certainly earn their space.

Leaf mould is most easily made with the leaves from deciduous trees, with those of beech, oak or hornbeam considered to be premium ingredients. Other leaves are fine, although some of the thicker, more leathery varieties, such as walnut leaves, should ideally be shredded to ensure that they rot down at the same rate as the finer, thinner ones.

Really tough evergreen leaves such as holly or eucalyptus are better stored separately; they, too, will rot down eventually, but may take a few years.

And finally, the needles from conifers can be used to make leaf mould, but should be collected into (another) separate pile. Pine and other needles are slow to rot down, but when they do, they tend to make a very acid mix that is particularly suitable for using on ericaceous plants such a rhododendrons, azaleas, camellias or even blueberries.

Arbor-vitae, ***Thuja***

Beech, ***Fagus***

Chapter 4

Everyday Garden Care

Is my garden too tidy?

If you've always been meticulous about tidying your garden at the end of the season, it can be disconcerting to hear that most of the wildlife you like to encourage would prefer things a little less… manicured. But if you're going to leave some messy corners, what kind of mess will actually help?

This depends on the size of your garden – those with small patio gardens may not have many options, although even on a patio you can leave a pile of empty pots and dried leaves in a tiny corner. If you prefer to do a big autumn clear-up when the soil is dry enough to walk on and you can get rid of any diseased foliage, you can still leave a corner or two overgrown and undisturbed. If you're lucky enough to have a slightly larger plot, you can create a mix of wild corners and more formal, kempt areas. And don't forget that some garden necessities, such as compost bins or layers of mulch, are mini-ecosystems, rich in potential food for birds, amphibians and insects.

Part of a larger whole

Rather than thinking of your garden as a separate entity, consider it as part of something much bigger. Nearby parks, street trees, rivers, canals and railway banks all contribute to the richness of the wildlife environment. Gardens can supplement this wider system by forming corridors between naturally wilder habitats. Help yours to play its part by adding any extras you can, such as bird or bat boxes.

Think of insects

Insects overwinter in the soil, in all their forms: larvae, pupae and adults. Giving them cover in the garden, and leaving the soil alone

Offer as broad a range of food and shelter as possible – such as this bat box – to attract plenty of different species.

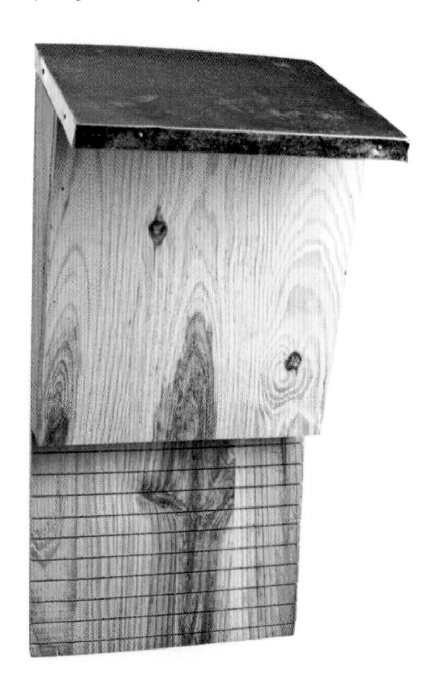

during the winter months, will ensure that they survive into spring. There's also the option of planting a cover crop of mustard or ryegrass on any larger areas of bare soil; this will be good for both the soil and the wildlife and can be dug in, while still green, before you sow again in spring.

A

It doesn't have to be all or nothing: think of your garden as a piece in the patchwork of landscape around you. Plenty of variety is the best thing for wildlife.

GOOD COVER, AND WHO USES IT

- Standing seed heads offer good food opportunities for seed-eating birds, while hollow stemmed plants offer shelter for lots of different species of insects and spiders.
- Ladybirds will collect together in clusters on dead foliage or plant stems.
- Worms will pull leaf and plant debris left on the surface down into the soil (and will thereby enrich it).
- Bricks, or upturned pots and saucers can offer cover for all kinds of insects, as well as for toads and newts.

Can I have a meadow instead of a lawn?

NATURAL FLOWER MEADOWS IN OPEN COUNTRYSIDE look so appealing that it's tempting to try to recreate one in your garden. But how feasible is it to grow your own – and how much maintenance does the 'natural' look call for?

Flower meadows come in two types – annual or perennial. The perennial meadows work best on poor soils because grasses won't compete with the wildflowers as much.

> Wildflower meadows aren't quite the no-work option you might assume. The perennial variety in particular is perhaps not for complete beginners, but if you have some gardening experience and are enthusiastic, it's well worth trying one out.

Growing an annual meadow

An annual flower meadow – in which the flowers are traditionally a mix of corn cockle, corn marigold, common poppy, cornflower and chamomile – calls for richer soil. Cultivated garden soil tends to be relatively rich, so a meadow is likely to do well without much additional work. It will look good over a single summer and may reseed itself, although the mixture tends to become less diverse over time. It's also easy to save seed from an annual flower meadow and do some of the reseeding yourself. Some commercially available wildflower seed mixes contain a wider selection of garden annuals and,

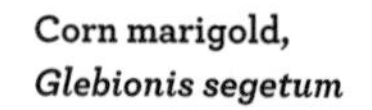

Corn marigold,
Glebionis segetum

Common poppy,
Papaver rhoeas

Chamomile,
Chamaemelum nobile

YOU CAN BUY YOUR MEADOW

Wildflower meadows are a business, and many suppliers offer ranges of different seeds, plug plants and pre-sown turf, tailored to a range of end effects. For a price, you can purchase almost any meadow you want, although you'll still have to prepare the ground to grow it in.

while not strictly mirroring the 'natural' mix of seeds, will result in an extremely colourful and pollinator-friendly display.

The perennial option

The conversion of a longstanding lawn to a perennial wildflower meadow means taking a long view – it can be several years before the balance of grass and wildflowers is right.

Some soils, especially clays, have such reserves of fertility that establishing a meadow can be a challenge because the grass will thrive and compete with the flowers. If your garden's soil is naturally rich but your heart is set on a meadow, it's worth considering the option of removing most of the topsoil and resowing, or using pre-sown wildflower turf, from which good results are almost guaranteed. A middle way, if you don't want the hard graft of removing the soil, is to spread an 8cm layer of coarse sand over the soil – this often proves to be a successful shortcut.

Corn cockle, *Agrostemma githago*

Three steps to a perennial meadow

If your soil is poor and light, you can create a perennial meadow out of an existing lawn. First, leave the turf to its own devices – don't kill the weeds or feed it. Mow weekly in the first year, removing the grass clippings. Second, let the grass grow long and flower in the following year. Lastly, after flowering, let the wildflowers that appear set seed, then mow and remove and compost all the grass. Repeat steps two and three every year.

A mix of the annuals that would grow naturally in a wildflower meadow; the flowers make a vivid impression against the green.

Q Can I have children and a lawn?

IT'S SAID THAT NOTHING IS HARDER on a lawn than children – except, possibly, for dogs. If you have the first, the second or both, is it feasible to have a good-looking lawn?

If your lawn is going to get quite a lot of use, choose a lawn grass that is tough and able to withstand heavy traffic – usually a mixture of perennial ryegrass, tall fescue, red fescue and a mix of other grasses known as browntop. Then aim to keep it at mid-length (2.5cm), rather than the super-short crew-cut that characterises the kind of lawn that boasts stripes. These coarser grasses grow quite quickly and will need mowing weekly; collect the clippings, too, to keep it looking as neat as possible.

A The cultivation of modern dwarf ryegrasses means that you can have a hard-wearing lawn, although it won't look like a bowling green. Alternatively, if you have the space, you could always separate the two – a play area for the children and a smooth lawn on which ball games aren't allowed.

Tall fescue,
Festuca arundinacea

WHAT ABOUT DOGS?

Careful lawn-owners have spent years trying to solve the problem of the scorched-looking areas of grass that mark their dogs' regular peeing spots, swearing by anything from dosing their dogs with cider vinegar to reducing the protein in their diet. Manufacturers have also produced patent remedies for lawn burn, which is caused by nitrogen in the dogs' pee. Opinions vary as to the degree to which any solutions work. Choosing the right grasses for your lawn wil help: US studies have singled out ryegrasses and fescues as the grasses least susceptible to urine-burn.

How can I make my lawn lush?

WHAT IF WHAT YOU REALLY WANT is a model greensward: a smooth, bowling-green expanse of bright emerald lawn – with stripes? Is it possible in an ordinary garden?

If you're prepared to put the work in, you can get a lush lawn through seasonal maintenance, using fertiliser and weedkiller, and mowing.

How to do it

A great lawn calls for seasonal treatments. In autumn, spike the lawn and rake out the 'thatch' (dead material); in late winter, treat it for moss. Feed and weed it in spring, water and feed as required in summer and mow regularly and tightly. You have to be ready to put in the effort (and the financial outlay on fertiliser and weedkiller), but the sleek results should repay the labour. If you prefer to garden organically, seaweed and sieved compost work just as well to feed and fertilise as their non-organic counterparts.

Most grass growth is in response to light, rather than warmth or watering, so shade is the element that will compromise a lawn the most. Cut back, as far as you can, any surrounding foliage that creates shadows on the lawn.

MOWING TIPS

A great lawn calls for a lot of mowing, so:

- Buy as large a capacity of mower as you can afford
- Consider a mulching mower that double-cuts the grass, creating fine shreds that are returned into the lawn rather than collected. Leaving the cuttings in situ results in the lawn needing less fertiliser in high summer.
- For a really low-labour solution, buy a robot mower and set it to wander about the lawn, snipping away, even when you're not at home.
- For stripes, you need a mower with a roller – there are plenty of options, both rotary and cylinder, available.

Should I be anxious about aphids?

AS GARDENERS, APHIDS ARE ALWAYS WITH US; in greater or lesser numbers, they show up every year, unwanted visitors to the garden party. But apart from being unattractive to look at and harmful to a plant in greater numbers, do they really do any serious damage?

If they're allowed to breed unchecked, aphids will eventually harm the plants they feed on. If they infest young plants they can infect them with viruses, and some aphid 'specialists', rosy apple aphids, for example, can damage the development of the plants they favour, as well as their crops. The honeydew they exude is soon covered with black mould. The best tactic is to practise control before they start to take over.

Aphids have a huge number of predators. Almost everyone in the garden wants to eat them. Ladybirds in particular are enthusiastic aphid eaters, are highly mobile and have an impressive ability to locate prey – so if the greenfly have turned up and started multiplying, the ladybirds will appear very soon afterwards.

Speedy breeders

If just a few aphids seem to have turned into an infestation overnight, a look at their breeding habits will soon explain why. They are capable of astounding population surges. Not only can they produce offspring

They're a feast for the ladybirds, but if they turn into an infestation, you may have to dispose of the aphids yourself.

without fertilisation, a process known as parthenogenesis, but during spring and summer they are also viviparous – meaning that they give birth to live young (skipping the egg-laying-and-subsequent-hatching stage). Even worse is the fact that every female aphid has a daughter inside her, and inside the daughter, even before birth, is another daughter – so every female represents three generations in one tiny package. This is called 'telescoping of the generations', and makes for extraordinarily efficient reproduction. In fact, the entomologist Stephen A. Marshall estimated that, given favourable conditions, and in the absence of any predators or diseases, a solitary aphid could produce 600 billion young in a single season. Thank heaven for ladybirds.

Aphids in winter

Aphids are flexible. In the warmer seasons, they are wingless and also viviparous. But it's all change when the weather gets colder. Some species overwinter as adults. But in others, dropping temperatures trigger the females to give birth to both males and females with wings. They then fly to new territories, where they mate and the females lay eggs on as-yet-untapped perennial plants. When spring comes, the eggs will hatch – and give birth to next year's first generation of wingless young.

TAKING MATTERS INTO YOUR OWN HANDS

The good news is that even if you have a year when there are so many aphids that the ladybirds and other predators can't keep up, there are plenty of other environmentally-friendly ways to control them. Check for early appearances on your daily walk round the garden so that you catch, and deal with, new groups early, before they take serious hold on a plant.

- If there are only a few, go back to basics: squash them by hand. Squeamish gardeners can wear gloves.
- If the plant they've appeared on is not very young or very tender, hose them off using a jet of water from the high-pressure setting on the hose.
- Where they have suddenly appeared in large numbers – and it does happen – use an approved insecticide. These are effective and if used as directed have no adverse effects on people, pets or the environment.

Q How can I fatten up a thin hedge?

A THICK, WELL-GROWN HEDGE makes a satisfying boundary in the garden: wildlife-friendly and more natural-looking than a fence. But too often a previously healthy hedge grows 'open' and thin at the base, giving it a much less appealing appearance and reducing its effectiveness as a barrier. How can you thicken it up again?

A

It depends on what kind of hedge it is – deciduous or evergreen. A very hard cutting-back is the traditional way to thicken up a hedge that has thinned at the base, and it works on most deciduous hedges. With the exceptions of yew and sometimes thuja, though, conifers can't be heavily cut back, so sometimes a more radical solution is called for.

▼ Successful hedges are dense and green all the way down to the ground. This hedge has avoided any thin, patchy areas at the base.

If a deciduous hedge has grown thin and leggy, pruning it, feeding it from the base and a generous layer of mulch can all help. If it has become very open and gappy at the base, though, it may need to be cut down to a height of around 30cm (a process known as renovation pruning) so that it can send up fresh growth. While deciduous hedges can recover from periods of neglect, conifers often won't. Apart from the two exceptions already mentioned – yew and thuja – they won't survive a radical cut-back, so if an evergreen hedge has become very scraggly and thin, the best option is often to replace it altogether.

HEDGELAYING: A TRADITIONAL CRAFT REBORN

In the past, agricultural land needed hedges that were strong enough to withstand livestock – and, if the livestock was a bull weighing more than 100kg, it meant that they had to be very strong indeed. Hedgelaying, or -layering, the technique that made them so impenetrable, is a kind of pruning, carried out in winter and early spring. Stems are cut almost all the way through with a curved billhook, then bent back into the body of the hedge at an angle. The work is skilled, and the resulting hedges are dense, wildlife-friendly and need only occasional maintenance. The craft of hedgelaying, which had almost died out in the latter half of the twentieth century, has recently been resurrected by a number of enthusiasts and societies, and laid hedges, while no longer widely used on agricultural land, are becoming increasingly popular in grander gardens because of the unparalleled shelter that they can offer nesting birds and other types of wildlife.

The angled, interconnected twigs and stems of 'laid' hedges create great habitats for insects, small mammals and nesting birds.

How to stop a hedge from becoming patchy

1 Plant it in autumn or winter, and clip it to a height of 45cm when it first goes in.
2 As it grows, keep the sides neatly clipped, but leave the top alone until it approaches its desired height.
3 Don't allow it to become top-heavy. It's best to shape the hedge so that the base is slightly broader than the width at the top. This means that the upper part of the hedge doesn't deprive the lower portion of light, which would cause it to stop it growing healthily.

Can I grow a hedge in a hurry?

USED TO MARK A GARDEN BOUNDARY, a well-maintained hedge can look more appealing than any fence or wall. But how patient do you have to be if you want to grow one from scratch – will it take years to see any results?

Traditionally, hedges are grown from 'whips' – that is, plants between 60 and 90cm high – but other heights can also be bought, although the larger the plant, the longer it takes to establish itself. And of course in terms of real growth, how fast a hedge progresses depends on what it is grown from. Hurdles made from woven willow or hazel are traditionally placed between growing hedgerows to offer privacy and support while the hedge grows – the hedge gradually covers up the hurdle, which will eventually rot away within the body of the hedge.

Small, glossy, regular leaves and fast growth make privet an ideal hedge plant; once grown, it's also tolerant of drought.

Slow and steady

Quick-growers may seem great at first, as they swiftly attain a 'proper' hedge height of 1.5m-plus, but they can become an ongoing nuisance because of what can seem like a near-constant need for pruning. Slow-growing hedge plants call for more patience, but at the end of the process, the hedge that results is low maintenance, needing far less ongoing care. Evergreen hedges are best planted in early autumn, and deciduous ones from mid-autumn to the end of winter. The former do offer the advantage of year-round privacy, while the latter are obviously less effective for screening between leaf-shedding season and the arrival of new foliage in spring.

A To some extent, how fast you can grow a hedge depends on how much you can afford to spend. At a price, it is even possible to buy an instant hedge, ready-grown in troughs, to be slid into position at its destination site.

TOP HEDGING PLANTS TO CONSIDER

Privet, *Ligustrum*
Pros: Semi-evergreen, tolerant, quick-growing and cheap.
Cons: Needs cutting twice a year; susceptible to honey fungus.

Leyland cypress, x *Cuprocyparis leylandii*
Pros: Very fast-growing (as much as 1m per year), pretty, green mossy effect.
Cons: Must be cut annually; bad reputation due to uncontrolled Leyland cypress hedges being allowed to grow rampantly, with oppressive results.

Portugal laurel, *Prunus lusitanica*
Pros: Evergreen, elegant, with a holly-like appearance, but much quicker growing (up to 60cm per year).
Cons: Must be cut at least annually, or even twice annually.

Leyland cypress, x *Cuprocyparis leylandii*, grows extremely quickly; it's great for growing a hedge in a hurry but must be regularly trimmed.

Arbor-vitae, *Thuja*
Pros: Evergreen, comparatively low maintenance.
Cons: Can grow straggly; needs an annual trim.

Beech, *Fagus*, or hornbeam, *Carpinus*
Pros: Both British natives make beautiful hedges and, although deciduous, retain their dead leaves very attractively in winter. Can be cut back without problems if they get out of hand.
Cons: Slow to grow.

Box, *Buxus*
Pros: The traditional selection for close-clipped formal hedging; makes a good-looking low hedge.
Cons: The box-tree caterpillar and box blight, a destructive fungus, have detracted from the plant's popularity in recent years. *Berberis* and *Pittosporum* are both viable alternatives that aren't disease-prone, and both of which can be clipped into neat, low hedges.

Q How can I fill garden gaps?

HOWEVER WELL YOU PLAN, seasonal gaps usually appear at some point in the gardening year – new plants fail to thrive, old ones die unexpectedly or a space that was filled with a plant that was strictly seasonal is left vacant when its season is over. What's the best way to plan for, and fill, the gaps?

A

Be prepared – grow extras to fill in for unexpected failures – or buy late in the season for a strong, but short-term, effect.

Before rushing to fill gaps, first check that you need to. In a crowded border, cutting back spent or failed plants releases others from competition, and you may find that existing plants expand to fill the space vacated much more quickly than you'd envisaged.

Gap or opportunity?

Sometimes, though, there's a noticeable gap that you'll want to fill fast. Garden centres and nurseries supply large, later-flowering plants, such as begonias, dahlias, fuchsias, penstemons and salvias. They may not be cheap – although, by July, many will be on sale or on special offer, but it's a chance to buy relatively large plants rather than the smaller examples you'll find earlier in the year. You need them to be large as the season is running out and small plants won't get big enough quickly enough to earn their keep visually in the remaining weeks. On the other hand, some annuals, raised from seed – for example, calendulas, nasturtiums and phacelias – are fast-growing enough to flower before the cold weather arrives. Once in place, be prepared to spot-water and feed the new arrivals to get them settled in as speedily as possible.

Match the filler plant to the location – crocosmias, penstemons and gauras will thrive in sun; for gaps in shadier corners, consider hostas or rodgersias.

Keep some pots in reserve: when you're growing for the garden, pot up some seedlings at the same time as you plant the others out into beds, so there are ready-grown spares for gaps when they arise.

Begonia hybrid
Begonia

How do I fix a neglected garden?

IF YOU HAVEN'T GARDENED REGULARLY for a long spell, you can reach a point at which your garden seems to need so much work that you don't know where to begin. Rather than worrying about it, it's best to make a realistic plan and start small.

Tidy up

Begin by clearing things up. This may feel like fiddling while Rome burns, but it will give you a better picture of the work that's really necessary and will help you to feel more in control. Cut back any extreme overgrowth and edges, and tidy any paths – this will have a disproportionately positive effect in relation to the amount of work involved. Take your rubbish bags to the local green waste site (don't leave them sitting around for days; the clearer your garden, the more straightforward the next stages will seem).

Look at what's left. Which jobs are essential? Do you have a problem with brambles? Or perennial weeds? Is the soil in decent condition? Is the plot already over-planted (in which case you may be thinking about taking things away, rather than adding them). Make a list of all the things that need doing, and finish each task before starting on the next. If you feel your motivation waning, give yourself a time frame. Tell yourself you'll complete one of the larger tasks every week, then don't go beyond this. This means that you won't get too tired or bored to stick to the overall plan.

BROWSE, DON'T BUY

The temptation is always to rush out and choose plants when you're engaged in a garden makeover, but don't buy anything until the garden is in better order. When you've cut back, weeded, conditioned the soil and taken out any plants you've decided that you don't like, or that aren't thriving, it's time to fill any gaps. In the meantime, you can window-shop from catalogues – but only buy when you're ready to plant.

A Divide the work into manageable tasks, and don't drive yourself crazy through constantly reviewing the whole and fretting about what a huge job it represents. Try to think of it as an opportunity to get the garden you really want, step by step.

Can anything stop slugs?

WE'VE ALL BEEN THERE. One day you have a newly planted border of flowers or a newly sprouted vegetable patch, the next morning you discover that it has been ravaged by slugs. Universal and unforgiving, they are perhaps the most dispiriting of garden pests.

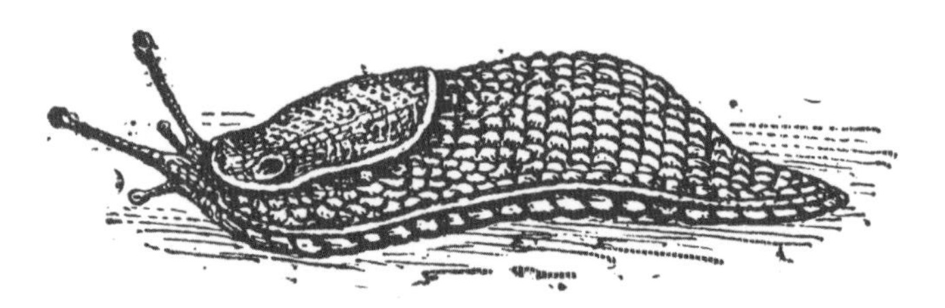

They may not look particularly threatening, but in the garden, slugs act as impressively – and dismayingly – efficient feeding machines.

Slugs live above and below ground, so they are just as much of a threat to roots and tubers as they are to leaves and flowers. Superlative feeding machines, they leave all-too-familiar irregular holes in many garden plants and silvery slime trails in their wake. What's more, providing temperatures are above 5°C, they will continue to feast on your garden plants thoughout the winter as well. The good news is an individual slug will only live for around a year. The bad news is that slugs are a hermaphrodite species, so each of those individual slugs can lay up to 300 eggs a year, at any time.

Fighting back

So, can anything stop slugs? Some gardeners swear by copper strips, eggshells, coffee grounds and the like to prevent slug ingress into borders, beds and pots. In truth, however, these are not terribly effective. Slug pellets and biological control in the form of nematodes can reduce a population temporarily and can help in spring or autumn (see p133). And while beer traps will send slugs to a boozy death, they need regular emptying, which is frankly disgusting.

Most effective is actual physical removal of the slugs in regular dawn or night-time patrols. Complete elimination of slugs from a garden is impossible, so it is wise to concentrate on protecting the most vulnerable plants: those newly planted or sown, or fresh new growth of herbaceous plants in spring. Where you dispose of them is another matter. Slugs and snails surreptitiously lobbed over the fence will often find their way back, as they leave scent trails in their slime. Placing them 20m away or more is necessary to deter them from returning. Another option, if you don't mind drowning them in a bucket of water first, is to add slugs to the compost heap.

A Put the pellets away and instead focus your energy on creating a healthy biosystem with natural predators – including yourself!

A HOLISTIC APPROACH

As with many garden pests, the cultivation of a healthy biosystem within the garden will help keep slugs in check. Encourage birds, hedgehogs, frogs, toads, slow worms and beetles that will feast on the slugs. Remove as many slug hiding places as possible and rake the soil to reveal the eggs. And remember that, while they can be annoying, slugs aren't all bad: they eat around twice their body weight every day and most of it is decomposing material, making them invaluable to compost heaps and the soil ecosystem in general.

Birds, such as blackbirds

Frogs and toads

Hedgehogs

Ground beetles

Q How can I attract more birds to the garden?

AT A TIME when there's a growing awareness of the decline in many species of small birds, the kind that used to be frequent and numerous visitors to most gardens, what's the best way to encourage more birds to visit the garden?

Fieldfares, *Turdus pilaris*, tend to feed in flocks, and a crowd may sometimes descend to feast on a favourite food, such as crab apples.

A

Shelter, food and water are the best inducements you can offer. Do your research into what species you might be able to attract, given your garden's location, and then look at what they prefer in terms of natural food sources.

With lower temperatures and fewer easy food sources, winter is the toughest time for birds. Keep a clean water source for them, whether it's a birdbath or simply a plastic bowl, and refill it regularly, remembering to break the ice on freezing days so that they can still have a drink. For food, apart from the ready-mixes of nuts and seeds, the fatballs and other delicacies you can buy for bird tables and feeders, consider making space for some of the plants that bear autumn and winter fruits and berries that birds particularly relish. You'll be repaid with the enjoyment of watching them feed.

Great globe thistle, *Echinops sphaerocephalus*

AUTUMN AND WINTER FORAGING – FOOD PLANTS FOR BIRDS

Firethorns, *Pyracantha* – autumn and winter berries.

Rowans, *Sorbus* – autumn and winter berries.

Hollies, *Ilex* – autumn and winter berries.

Ivies, *Hedera* – autumn and winter berries (eaten by birds in midwinter).

Cotoneasters, *Cotoneaster* – autumn and winter berries.

Honeysuckles, *Lonicera* – autumn berries.

Crab apples, *Malus* – autumn and winter fruits (birds prefer the smaller fruits cropped by species such as *Malus floribunda* and *Malus sargentii*).

Like most berries, those on rowan trees are ideal bird food in winter, as they are high in calories and packed with digestible sugars.

Globe thistles, *Echinops* – autumn seedheads.

Plume thistle, *Cirsium rivulare* – autumn seedheads.

Blue-berried honeysuckle, *Lonicera caerulea*

Why do my plants keep blowing over?

PLANTS NEED SUPPORT of different types and at different times – and this support needs to be surprisingly strong to do an effective job. There's a huge range of possibilities available with plenty of options to work for every type of plant and situation.

If your plants are blowing over, you're either using the wrong type of supports for the plants, or the supports themselves aren't sturdy enough. Review what's available and choose something tough enough to keep the plants upright, even in very blowy or rainy weather.

Border perennials often need help to stop them flopping or drooping – and you can prepare for this by putting in the right supports early in the season, so that subsequent plant growth will make them less noticeable. Possibly the most challenging are the plants with individual flower spikes, such as lupins or delphiniums, or those with very large and heavy flower heads on less substantial stems, such as peonies.

Pick the right type

Options include straightforward upright stakes (effective, not easy to disguise), metal ring or half-ring supports (sunk into the ground early, so the plant can grow up through its support and rest on it when it has grown large enough) or stake-and-netting combinations. Many come

Many plants with spires of flowers, such as lupins, need support, but it must be unobtrusive or it can spoil their overall effect.

NATURAL SUPPORT FOR PLANTS

Perhaps the most visually appealing supports you can use are homemade 'cages' woven from peasticks. Named because they were originally deployed to support climbing pea plants, peasticks are the flexible, twiggy shoots from young birch or coppiced hazel trees. A pencil-width or a little thicker, they are pliable enough to weave into a tailor-made support and create cages over young plants to keep them supported as they grow through. If you don't have a hazel wood nearby to collect coppice leavings from, peasticks can be bought in bundles at many nurseries, or purchased online.

The classic way of making a cage from peasticks is to surround the young plant with a circle of sticks, then bend them down towards the centre and wind them around each other where they overlap. To be an effective support, the resulting cage shape should be between half and two-thirds of the height of the fully grown plant. As you become used to weaving them together, you can get creative and make different shapes and sizes to suit different plants. Visually, peastick supports blend in better than any of the other options, and they work very well. Their only real drawback is that they need to be replaced annually.

'Cages' woven from peasticks offer natural-looking, effective support for a range of plants, from peonies and asters to dahlias.

in green or black, and black can sometimes, surprisingly, be less conspicuous than green. Plants supported by stakes should be tied with twine, but in a figure of eight configuration, which prevents the stem of the plant from rubbing against the support.

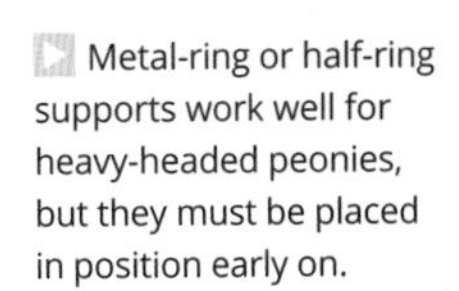

Metal-ring or half-ring supports work well for heavy-headed peonies, but they must be placed in position early on.

How can I spot honey fungus?

HONEY FUNGUS IS DREADED by gardeners – often it isn't spotted until it's too late for the plant it has attacked, and it's almost always fatal. It's found in North America, the UK and mainland Europe, and many other parts of the world, too. But how can you tell when one of your garden trees or plants has honey fungus, and can you avoid it infecting others?

If foliage turns yellow and starts to die, and new leaves are smaller and paler than they were last year, these aren't good signs. But when the bark at the base of the stem or trunk also cracks and begins to 'bleed', and the bark separates from the wood, revealing a white fungus with a strong 'mushroom' smell, the diagnosis is pretty much certain. It's honey fungus: nothing else looks or smells quite like it.

Honey fungus, *Armillaria*

Wet-looking splits and cracks in the bark at the base of this birch tree are some of the classic symptoms of honey-fungus infection.

Honey fungus belongs to the *Armilllaria* genus. There are numerous forms, but not all do damage in gardens. Although seven types are found in the UK, for instance, most of the damage to living plants is attributable to only two of these. It attacks the roots of trees and woody and herbaceous perennials and travels underground to infect others. In autumn, the honey-coloured toadstools for which

WHICH PLANTS ARE SUSCEPTIBLE?

Certain plants seem especially prone to honey fungus, while others are less susceptible. You can find comprehensive lists of both online at the RHS website.

The trees that seem to fall victim to the fungus most often include maples, willows, beeches, birches, walnut trees and rowans. Viburnums, roses, privets and rhododendrons are also among the plants that are more susceptible to honey fungus.

the condition is named may grow at the base of the affected plant. Late in the progress of the disease, you may find rope-like rhizomorphs under the bark or, less commonly, in the soil around the plant. Yellow-orange or brown at first, they will eventually turn black.

HONEY MONSTER

The largest living organism on Earth is believed to be a single honey fungus mass – *Armillaria ostoyae* – found in the Blue Mountains in Oregon, USA. When it was examined in 2014, it was found to be an enormous 3.8km across.

How to beat it

The way to eliminate honey fungus is to dig out the roots of the affected plant – very thoroughly – and burn them. Some gardeners place a vertical barrier of plastic sheeting in the soil to block off the area where the roots were, to ensure that any fungus left in the soil can't spread to other plants, but this is quite a challenging task for any novice gardeners. If a plant has been killed by honey fungus, wait a year before planting again in the same spot, and, when you do, choose a non-susceptible plant as a replacement.

Honey fungus travels to new 'host' plants via ropy rhizomorphs, which are found around the roots of those already infected.

Q How can I make my garden look more interesting in winter?

WINTER IS WHEN THE GARDEN RETREATS into itself, a time for plants to ration energy and prepare for the seasons ahead. For the gardener, this can also be true. While a lot of the season's work can be done from an armchair, winter gardens don't have to be dull.

One of the pleasures of having a garden is that it provides something nice to look at out of the window, and there's no reason why this can't be the case all year around. Indeed, all RHS gardens have winter bedding and permanent winter garden areas, and other gardens with dedicated winter gardens include Cambridge Botanic Gardens, Sir Harold Hillier Gardens and Anglesey Abbey.

Creating interest in the garden during winter is all about making the most of what's available in the season: leaves and flowers are in short supply, so the emphasis is on structure. Many of the best winter gardens include hedges and topiary, which look especially good in relief against frost and snow. Not being too hasty to cut back perennials can also make a winter garden less boring. First, the seedheads and old stems of various grasses and perennials, such as *Miscanthus*, *Eryngium*, *Rudbeckia* and *Echinacea* are attractive in their own right, especially when frosted. Second, they will also attract birds to feed on them. If the garden is not sufficiently interesting, adding bird feeders will also provide plenty to look at in the winter months.

A

Careful planting will provide interest throughout the winter months without forcing you out into the cold.

Winter flowers

Plants that do flower in winter are often heavily scented. Planted near doorways and paths where their flowers and scent can be most easily appreciated, shrubs such as daphnes (*Daphne*), Christmas box (*Sarcococca*), witch hazel (*Hammamelis*) and early stachyurus (*Stachyurus praecox*) add interest if not massive blooms. Showy winter flowers are generally restricted to bedding plants and bulbs. These are best displayed in containers near the house, although large drifts of snowdrops will herald the spring in late winter. Plant amongst the black-leaved grass *Ophiopogon planiscapus* 'Nigrescens' for a modern, monochrome effect.

FIVE TO TRY

Trees and shrubs with interesting or coloured bark will come into their own in winter. These include *Acer* species, such as box elder (*A. negundo*), and Père David's maple (*A. davidii*) or paperbark maple (*A. griseum*), as well as Corkscrew hazel (*Corylus avellana* 'Contorta') and white birch, such as West Himalayan birch (*Betula utilis* var. *jacquemontii*).

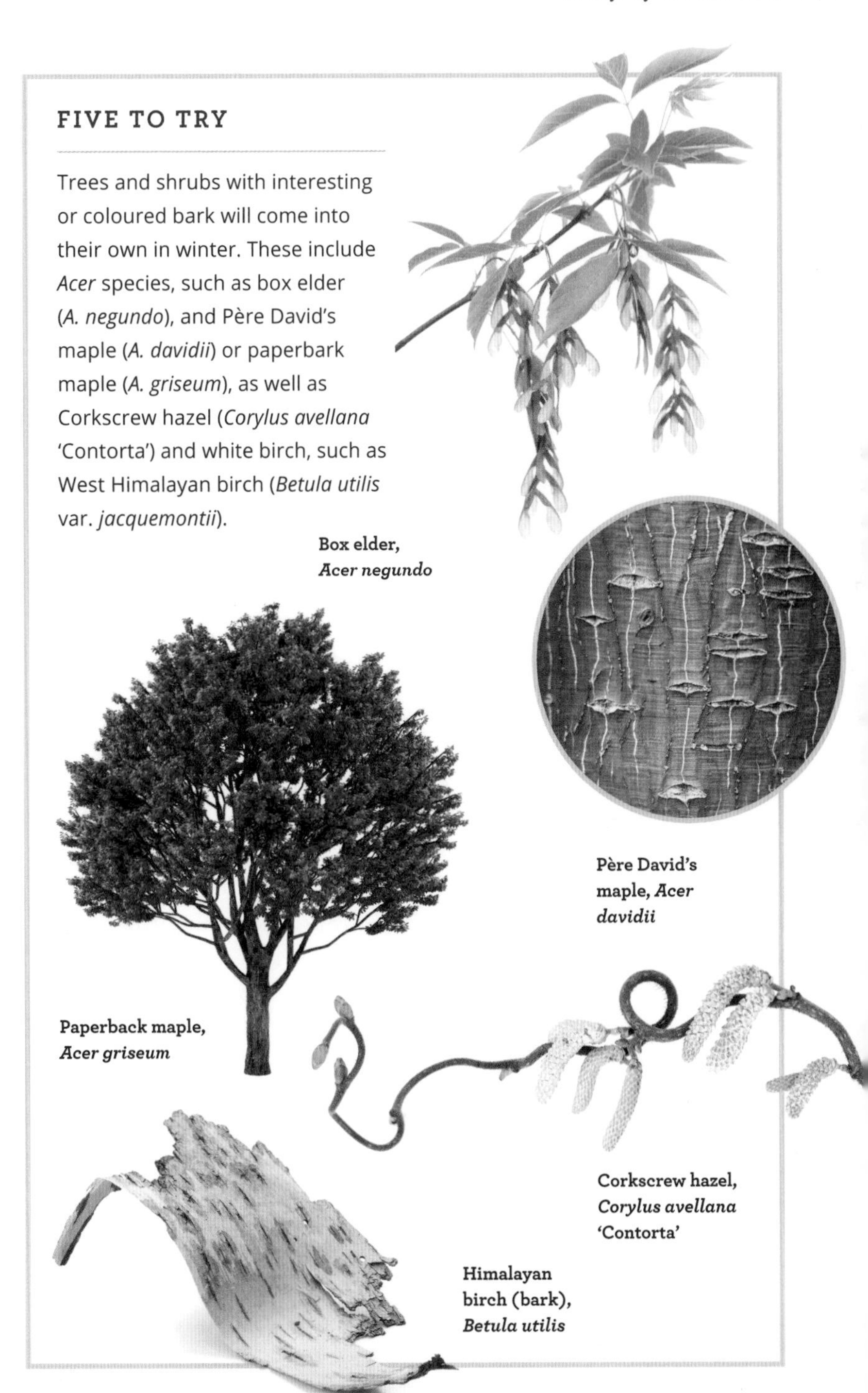

Box elder, *Acer negundo*

Père David's maple, *Acer davidii*

Paperback maple, *Acer griseum*

Corkscrew hazel, *Corylus avellana* 'Contorta'

Himalayan birch (bark), *Betula utilis*

Q How can I plan ahead?

Somehow, despite an enthusiasm for gardening books and magazines and all the inspiration they offer, it can seem as though you're always playing catch-up in the garden, without any certainty about what the next season is going to bring, or how best to deal with its opportunities. What's the best way to plan ahead?

A

To plan ahead, you have to... plan. This can mean mapping out your plot on to graph paper and filling in areas for each season, or thinking long-term, about the space your plants will need once fully grown.

Once it's started, the gardening year never really stops. In winter, you have some downtime to plan the seasons, and it makes sense to think ahead for the whole year, rather than just the next three our four months.

One year = four seasons...

The first part of the year tends to be easy, because it's all about plants getting going rather than finishing. But gardening happens on a roll, so you need to reconcile yourself to the idea that, as first the tulips, then the alliums come to an end in late spring, for example, the high-summer perennials will be starting to leap into action – there will always be something that's past its best that will need tidying and clearing, just as there will always (or at least until the very end of the gardening year) be something that is reaching its peak.

Plans usually depend on perennials for their essential framework, with gaps filled in with annuals or tender perennials, such as gazanias or verbenas, grown as annuals. Unless your garden is tiny, it is worth plotting it out on graph paper, copying the plan four times, and filling it in with plants for each season, looking at what space you have to spare, and how the various perennials will fit in with each other.

Treasure flower,
Gazania rigens

... or three, anyway

If your garden is small, there may not be much space for winter interest if you want to fit in plenty of plants that will be at their best in spring and summer, when you spend more time outside. If that's the case, aim to add some winter appeal with plants such as snowdrops, aconites, dwarf irises and hellebores, particularly in shady corners, so as not to compromise the later seasons.

Early bulbous iris, *Iris reticulata*

KEEP A PHOTOGRAPHIC RECORD

Apart from a planting plan on paper, one of the best aids you can have for the next year is plenty of photographs of past years. Take your phone out whenever you have a spare minute or two in the garden and keep a record of the things that you were pleased with as well as those that were less successful. This will mean that you can repeat and build on your successes (the wonderful mixture of an acid-green euphorbia and some almost-black tulips, say), and avoid duplicating any failures – risking another rose bush in the corner that a previous rose bush just didn't seem to like, for example.

Q Why did my plant die after pruning?

SHORTLY AFTER IT WAS PRUNED, a favourite plant died. Did the pruning kill it, or was the timing a coincidence?

A It depends on the plant. If they're pruned at the wrong season, plants can be left with too few leaves to keep the roots alive in winter. Evergreens such as holly or rhododendron, for example, do not respond well to pruning at any time except spring. Most deciduous plants are much more obliging. Although there are, inevitably, a few plants – broom and ceanothus among them – that don't like being pruned at all.

Californian lilac, *Ceanothus*

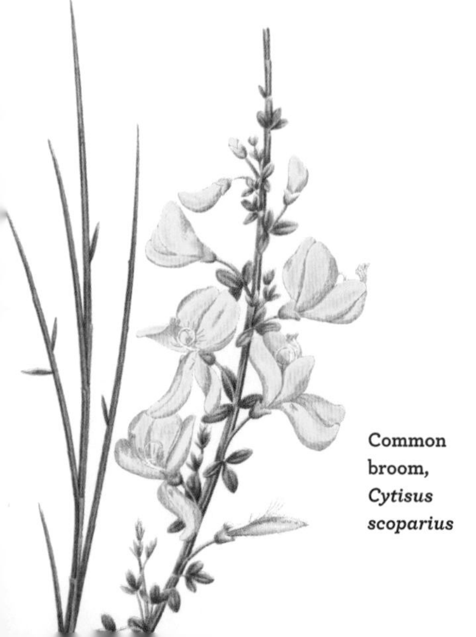

Common broom, *Cytisus scoparius*

Actually, it's relatively hard to kill a plant through pruning, although it's quite easy to lose a year's flowers (and to spoil its shape). If a plant died after pruning, the likelihood is that you finished it off when it was ailing anyway, rather than flat-out murdered it. On the other hand, a sickly plant may bounce back after pruning – most probably because the removal of poor-quality top growth

The key to successful pruning is getting your timing right. Forsythias, for example, must be pruned as soon as their flowers have faded.

relieved the roots from the strain of sustaining it, and they produced healthy new growth as a reaction to this strain being lifted. If you have a plant in extremis, pruning is worth a try, on the kill-or-cure principle.

Plants generally have a natural tendency, when possible, to maintain their root–shoot balance. This means that if some of the 'shoot' – top growth – is removed, the plant will try to grow some more in order to feed its roots. If it doesn't have enough resources to do this, it will retrench and sacrifice some roots – and the process can lead to a gradual decline in the plant's health and could eventually kill it. You can stage pruning over several years to avoid overdoing it.

A relatively common mistake can lose you a year's flowers in certain cases. Trees and shrubs that bloom in spring – *Forsythia* and *Philadelphus*, for example – will have made the following year's flower buds before they lose their leaves and become dormant in winter. This means that the only time you can prune them is pretty much directly after they flower, or you're at risk of pruning away next year's flowers.

RUINING A RELATIONSHIP

Sometimes pruning can have the less-than-obvious effect of causing a plant's roots to rot. Why? Because it caused the plant to reduce the quantity of sugars that it was secreting around its roots – and it was these sugars that were nourishing beneficial and protective microbes in the soil. Without the 'food' from the roots, the microbes died, and without their protection, the plant fell prey to opportunistic root rot or honey fungus.

Q Can I stop rampant ivy?

IVY IS GREAT FOR WILDLIFE, and can make an excellent and attractive cover for ugly surfaces – breeze-block walls, for example. Its downside for the gardener, though, is its extremely vigorous habit. Once introduced into a garden, how easy will it be to control?

A While it's true that some ivies grow extraordinarily fast, they won't take over if you don't give them the chance. This means that you have to be prepared to fit in some regular maintenance before your ivy has the chance to run amok.

Common ivy, *Hedera helix*, sends out little 'feelers' of roots along the entire length of its stems, which attach to any accessible surface.

Ivy berries offer a calorie-rich source of food for wild birds in the late winter, when other options are more scarce.

Ivy sends out many shoots, which can travel far and fast, both vertically and horizontally. Some species support themselves by putting out aerial roots that grip to surfaces as they grow. Be ready to prune it – hard – before it gets out of hand. Whether they're running along the ground or up a wall, the individual stems need to be pulled out from the growing end and traced back to the main stem before cutting (and all the small roots, where they're clinging to the surface or growing down into the ground, also need to be pulled out).

Although ivy is an opportunist, it's not a parasite – it takes advantage of trees that aren't in the best of health. With its very strong growth (it can reach heights of 30m), and the fact that it is not subject to many pests or diseases, ivy can smother a tree on which it grows. But in fact it is only using the tree to reach more light, and it has its own root system: it isn't feeding off the tree.

A GOOD ALL-ROUNDER

The benefits ivy offers to wildlife make an exceptionally long list. Here are just some of them:

- In spring, ivy makes good cover for nesting birds.
- The black berries stay on the plant from November to April. They are extraordinarily high in energy – weight for weight, they're packed with almost as many calories as a chocolate bar. Birds usually start to eat them in December or January, when earlier crops, from rowan trees and hawthorn bushes, are over.
- The dense foliage offers a shelter for hibernating butterflies.
- The flowers are a valuable late nectar source for a wide variety of insects. The UK's common ivy even has a dedicated bee, the ivy bee, *Colletes hederae*, which feeds exclusively from its flowers.

Red admiral butterflies, *Vanessa atalanta*, increasingly overwinter in Britain, and use ivy to roost in during colder or wetter weather.

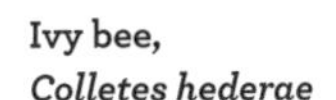

Ivy bee, *Colletes hederae*

Q Do I have to garden every day?

SOMETIMES GARDENING CAN FEEL like a never-ending chore. There's never a moment when you can say that it's finished: however good you get one bed or corner looking, there's always something that needs taking care of in another area. If you want to be a real gardener, do you have to garden every single day?

A

Gardening – unless it's your day job – is a voluntary activity. You can do it or not, as you please. But there's also something compulsive about it. Even people who have come reluctantly to it, having acquired or inherited a garden they weren't initially very interested in, often find that they've caught the gardening bug after a season or two.

Even if you don't intend to get your hands dirty, it's a good idea to get into the habit of walking round the garden every day. It lets you see what's happening, to spot what day-to-day changes are occurring and to look at what's doing well (and what isn't), what might be getting a bit out of hand and so on.

Everyday rewards

Gradually, it gets easier to enjoy the way in which the garden is in a constant state of flux, rather than resenting it. And with tiny changes happening every day, almost

The more enthusiastic you become about your garden, the more inclined you'll be to do one or two small, quick jobs daily.

FEEL-GOOD GARDENING

Of course, gardening is not just good for the garden; it's healthy for you, too. Keen gardeners have always been aware of the mental and physical benefits of their favoured pasttime, but these benefits have now become the focus of various studies into public health. In 2016, the King's Fund report was published in the UK. This looked at the impact that gardens and gardening has on individuals, but also at the wider effects of gardening in all sorts of other contexts – gardens owned by groups, or in public ownership were considered in terms of their broad social benefit. Unsurprisingly – to gardeners, at least – the conclusions drawn were that even experiencing time in the garden had overwhelmingly positive effects on feelings of well-being; it was recommended that practical gardening become part of a wide range of initiatives for improving public health. It had been proved: gardening is good for you.

invariably you start to want to engage with it. You don't have to start large: even a quarter of an hour spent doing one or two small tasks can make a difference, and will make the occasional big blitz seem less daunting.

Necessities

Watering (of containers, at least) needs to be done regularly, and certainly daily in warmer weather. Weeding, too, is best done regularly or a small (and, to be honest, dull) job turns into a large one. But the typical garden has a whole range of tasks that need completing, from the quick to the time-consuming. The 'every day' ten minutes can deal with most of the quick tasks.

Containers will invariably need regular watering; unless rainfall is torrential, pots don't usually catch enough to keep plants well irrigated.

Q Can I stop mosquitoes breeding in my water butt?

IN THEORY, mosquitoes shouldn't be able to breed in a water butt with a tight-fitting lid. But somehow it happens – how are they getting in, and how can the larvae be stopped from going on to make the next generation?

There's a huge amount of online complaint about this problem – which seems to happen more often than you might think. And there are some surprising solutions suggested, too: among them, the suggestion that you import some goldfish (enthusiastic predators of mos-quitoes, in all their forms) for a few days' water-butt holiday.

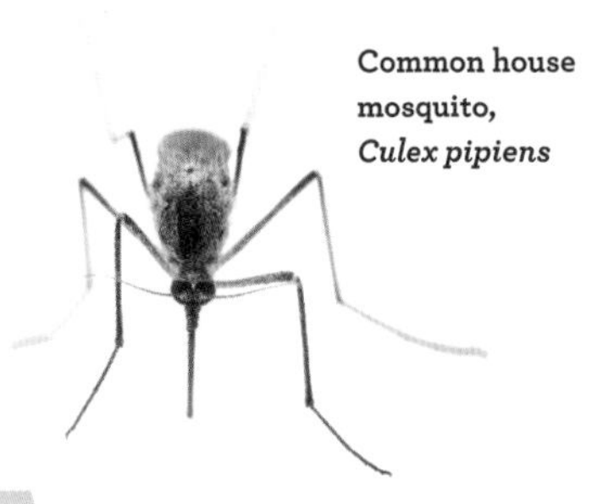

Common house mosquito, *Culex pipiens*

A It's a surprisingly common problem: it's possible the mosquitoes have found an entry point in some tiny space around the joint between the downpipe and the water butt, or that there's some other minute nook or cranny admitting them. Once the adults have laid their eggs and the larvae have hatched, however, it's easy enough to stop them: just oil the water surface so the larvae can't breathe.

A substantial plastic water butt looks impregnable, but mosquitoes can make their way in through even the smallest crevice or gap.

There's no need to trouble the goldfish, though. Mosquito larvae hang just under the surface of still or stagnant water, but breathe above it through a 'siphon' that extends above the surface. If you create a very thin layer of vegetable oil on the surface of the water in the butt, the larva's siphon won't be able to break through. Make sure that the layer is very thin – you don't want to create an oil slick. If you add a couple of teaspoons of oil to a small sprayer full of water, shake it well, and then spray all around the water surface in the butt, that should do the trick.

A very fine film of oil is enough to prevent mosquito larvae from breathing; dilute it substantially first.

TAKE CARE OF YOUR BUTT

The question of breeding mosquitoes raises the whole issue of looking after your water butt or butts (many gardeners value rainwater to the point that they have more than one). If they're left unmaintained for too long, they can get very murky and algae-ridden, so put an annual service on your gardening to-do list (if you're cleaning the gutters, that's the time to include the water butt in the job).

Summer's a good time to do it, as you'll be using plenty of water in the garden, and it's an easier job in warm weather. Take a week or two to empty the butt and remove the piping connection. Turn the butt on its side and wash it out; scrub thoroughly to remove all algae, rotted plant matter, moss and other debris. Clean the connection pipe, too, and check that it fits tightly (could this be where those mosquitoes have been finding their way in?) If a lot of plant material seems to have made its way to the bottom of the butt since you last cleaned it, one option is to invest in a water butt filter with an overflow, which will filter out plant matter before it gets into the butt.

Chapter 5

The Garden and Beyond

Q What can I do with a tiny garden?

A VERY SMALL GARDEN can seem at first to offer fewer possibilities to the gardener – there isn't space for a huge variety of plants, and if the site is enclosed, it may feel claustrophobic. Turn this thought around and envisage a minute jewel-box of a garden instead, and you will start to see its potential.

A Think in terms of the opportunity a tiny garden offers – it won't demand long hours of maintenance, so you can keep it looking wonderful. Use vertical spaces, using plant stands and training climbers to grow up walls. You could also try painting the walls.

The garden will seem more exciting if you can't see all of it at once. Add one or two levels with a couple of steps down, or an arch with climbers, even if it only leads to a teeny alternative space.

If there's a wall at the 'far' end of the garden, painting it blue will elongate the space. Alternatively, place a large framed mirror on it, to double the depth of the garden. Put some solid stickers on the surface, though (any type will work). These will discourage birds from trying to fly through the mirror. If you're creative and like a challenge, a piece of painted *trompe l'oeil* on a small garden wall – an extension of garden plants, or an exotic landscape – will look terrific surrounded by real plants growing onto and over it. You need to make use of all the space available, and this includes vertical space: train climbers to grow thickly over the enclosing walls and fences to give the most verdant impression possible. You could also introduce additional vertical surfaces to a small garden by putting in a structure such as a pergola.

How to fill the space

Consider brick or paving with as many beds and borders as you can fit in for growing space. Pots are great in small gardens, arranged in groups, and can be planted successively as the seasons change to look their best all the time. Consider investing in a plant stand, too, so that you can place pots at several levels without using up much ground area. Pot hangers are also an affective use of any spare wall space.

What can I do with a steep garden?

ASTONISHING GARDENS can be created with terraces, but depending on how steep the slope and how large the garden in question, constructing the terraces can be hard work – and expensive, if you need to employ outside help.

Why not create a wildflower bank? Begin by deciding how much labour, and cash, you are prepared to spend on your steep 'garden'. Are you prepared to hire an engineer and a digger, or do you feel more able to go with what you've got, plant it up to its best advantage, and mow some narrow paths to make it easier to navigate (scrambling up steep slopes is tiring).

Steep banks laid to grass are challenging-to-impossible to mow closely, so a better (and also more attractive) option is to keep them as long meadow grass with flowers. If the site is very steep, wildflower turf or soil matting ready-sown with seeds may be easier than sowing from seed. A wildflower bank will need cutting once or twice a year, a job that is safely – though slowly – done with a strimmer.

Very steep slopes must be terraced to be feasible as gardens that are cultivated in the usual sense. If it is possible to create some flat areas, you can garden on them, but in extreme cases, you may need the help of a structural engineer to make walls a metre or so high to retain the soil. If that's too ambitious, look at the easier option of turning the slope into a wildflower meadow with mown walkways.

Q What can I do with a huge garden?

IT MAY SEEM DAUNTING, but other gardeners will envy you – most have to choose between vegetable plot, flower garden, seating area, lawn and so on. If you've time to maintain them, you're in the position to have most, if not all, these features.

A

When planning, begin with the area immediately around the house and work gradually outwards. Try and think in terms of different 'rooms' for different functions.

Unless you're in the unusual position of creating a huge garden from scratch (in which case, you will probably need some professional help), it's best to watch the garden through one complete cycle of the seasons before deciding what to change and what to keep. Photograph all aspects of the garden throughout each season, including taking views from an upstairs window if possible, to remind you of what looked good at different points of the year.

When the year is up, use the garden designer's approach of thinking of the garden in terms of different 'rooms'. If large areas are very open, consider hedge-dividers to create self-contained spaces that all have different functions – the vegetable garden, the dedicated children's area, the sitting-out space and so on.

For huge gardens, grass is one of the most practical ways to manage a really large space. Although mowing can be time-consuming, you don't have to maintain it with a short-back-and-sides cut; you could opt for a wilder effect and hire a flail mower twice a year. This cuts longer grass, leaving it with a rough-cut, rural look.

CONSIDER AN ORCHARD

Once a staple in the gardens of even moderately sized houses, the home orchard is now sadly rare. But an orchard is one of the best ways to help the environment, particularly if you look beyond apples and pears to all the rarely grown fruit and nut trees, such as medlars, quince, filberts, cobnuts and mulberries. Drifts of snowdrops and daffodils are beautiful growing in rough-cut grass under trees, and orchards are ideal for free-ranging hens.

Is fake grass forgivable?

ERSATZ GRASS, made from sturdy plastic, has traditionally been considered a no-no for any 'real' garden. If, however, you have a small space, want the look of a lawn, but don't have time to maintain it, it's proving an increasingly popular option. So what are the pros and cons?

Modern artificial grass looks more convincing than it used to, and is seen in plenty of small-town gardens, and gardens that are used by children. The bad news is that it's not an environmentally friendly choice.

It depends what you mean by forgivable. Fake grass needs careful laying – competent installation makes a big difference both to the end effect and to its longevity. Although it doesn't need watering or fertilising, it does require regular washing and can attract moss and weeds.

There's a cost to the environment in its production – it's made out of polymer plastic. And, as even the good-quality varieties look scruffy and need replacing after a couple of years of heavy use, there's another cost in its disposal: it creates non-biodegradable and non-recyclable waste polyethylene and polymers. Fake grass is relatively easy to remove, though, and the soil underneath will then recover more readily than concreted–over earth.

Fake grass is made like carpet: plastic strands are knotted into a backing, with an underlayer that keeps the 'blades' upright.

THE PAVING ALTERNATIVE

It doesn't create 'lawn', but one low-maintenance alternative that is friendlier to wildlife is to pave over the area, leaving some slab-sized gaps that can be filled with good-quality soil and planted up. The resulting mini-beds won't need much maintenance, are less deadening for their surroundings than artificial grass and can look very good. Excess rain will drain easily in the gaps, too, which is a definite plus when the ever-increasing number of cement yards is making the build-up of rainwater run-off a problem in many urban and suburban areas.

Q Why don't frogs like my pond?

WHEN YOU MADE YOUR WILDLIFE POND, you expected the wildlife simply to turn up and populate it. But a year or two on, you haven't seen any signs of frogs, let alone a newt. Why doesn't your pond seem to attract wildlife and how can you make it more appealing?

A

The pond is probably a haven for plenty of tiny animals – but it may be on a smaller scale than you hoped for. To entice amphibians, make sure the pond has shallow edges, place some stones in the pond's centre and check that the pond is between 60 and 100cm deep.

If you want to attract frogs, first you'll need to make sure your pond is offering the basics.

Things to check

Amphibians like shallow edges, to help them get in and out of the water. If your pond has steep sides, create a wooden ramp and prop it at one edge of the pond to make it easier for them. Create some shelter with vegetation that comes up to the water's edge on one side of the pond. Frogs spend plenty of time out of the water, feeding, so they need to feel safe around the edges of the pond, as well as in it.

Put some stones (or encourage some clumps of vegetation) in the middle of the pond to offer a place to rest out of the water, but from which it will be easy to jump back in again if danger threatens. Try to avoid adding even, flat surfaces such as paving stones – these heat up a lot in strong sun, so can become too hot for amphibians to rest on.

▼ The ideal habitat for frogs includes plenty of shelter around the water's edges, as well as easy-access routes in and out of the pond.

Check that your pond is the optimum depth for wildlife – ideally, it should be between 60–100cm deep.

Try to make sure you aren't tidying up too rigorously: let the garden stay just a little messy. In terms of small-animal shelter, the pond shouldn't be a single oasis in the middle of a desert. There should be other places to shelter – a heap of logs, some unraked leaves and so on.

Things to be aware of

Wait for frogs to arrive of their own free will – don't transport frogspawn or adult frogs from another site. It's frowned on by wildlife experts who stress the possibility that you'll move diseases around by shifting them.

Fish view frogspawn as a delicacy, and they love tadpoles, too. So if you hope to attract frogs, don't put any fish in the pond.

POND CLEANING

Barley straw can be used to keep a pond algae-free, but must be kept floating and replaced every few months – special containers that will keep it floating can be purchased in nurseries and garden centres. Add it to the pond in spring and remove it around six months later. Adding a few lavender stems to the straw will make it even more effective.

Ponds need clearing out every so often to avoid them getting clogged with excessive weeds and mud. If you're planning a clear-out, aim to do it in early autumn – frogs sometimes like to hibernate in the mud at the bottom of the pond, and will usually turn up in later autumn, so delaying your annual clean up may mean disturbing the wildlife that you've been trying to attract.

▶ Barley straw contained in a pair of tights works as an easy, natural way to keep pond water algae-free, but may take up to six weeks to have an effect.

Q Is it easy to create a green wall?

GREEN WALLS – walls that are densely and evenly covered from the base all the way up to the top with plants – are increasingly featured, looking verdant and striking, in gardening magazines and articles. But how easy is it, first, to create a green wall, and, second, to keep it looking its best?

A

'Green wall' is a term that describes a wall with pockets or growing areas mounted on it to allow a mix of different plants to take hold at different points on the wall's face. Making one is a little more complicated than simply covering a wall with climbers, grown from the base, but should be well within the capabilities of a keen home gardener.

If you want a garden-on-a-wall, with many different plants, start by looking at the various kits available, which are basically frames holding pocket containers. The containers are filled with soil, or another planting medium, and planted up before the frame is mounted onto the wall. They need to be sturdily fitted and the containers regularly maintained. In particular, any 'casualties' need to be removed and replaced, or the wall will develop 'holes' in its surface – and most of a green wall's charm is in its solid-foliage look.

Large-scale systems are available, too, but need professional installation – they usually involve the delivery of pre-planted panels – and sometimes professional maintenance, too.

▼ In time, the separate plants of a green wall will grow together and intermesh, creating the impression of a solid green surface.

Pros and cons

Environmentally, a green wall will offer valuable extra habitat for insects and will also insulate the wall it covers, keeping it cooler in summer and warmer in winter. And if there is no soil at the base of the wall to allow you to grow climbers or shrubs, it may be an easier – and is certainly a more instant – option than trying to grow climbers in containers and keep them thriving.

The only obvious 'con' of a green wall is the amount of work it takes to install, although if this is done carefully, it should require only general maintenance afterwards.

However, whether you opt for DIY or a professionally installed system, it's always sensible to consider how you're going to water the plants. An irrigation drip system, which automatically waters the wall, is a great idea and makes watering simple. Alternatively, a hose can be used, with the help of a ladder to access the hard-to-reach spots – just be careful not to damage the greenery.

WHICH ARE THE BEST PLANTS TO CHOOSE FOR A GREEN WALL?

Generally, plants with a short and dense natural habit will do best. Three good examples to try are:

Bugle, *Ajuga reptans*. This has deep-green leaves that form a dense mat, with 15cm spires of blue flowers, like tiny snapdragons, in late spring.

Alum roots, *Heuchera*. Heucheras have foliage in a range of different colours, from deep bronze or near-black to almost pastel shades. 'Key Lime Pie' is a particularly attractive example: it is bright green with paler marbling on the leaves

Japanese sedge, *Carex oshimensis* 'Evergold'. This is a tough perennial clump-forming sedge with handsome yellow-striped, strap-shaped leaves.

Many herbs, including oregano, marjoram and various thymes will also thrive as a fragrant part of a green wall.

Bugle, *Ajuga reptans*, is ideal for green walls; its foliage rambles and 'mats' readily, and it has blue flower spikes that stand clear of its leaves.

Q How can I blast my neighbour's bamboo?

Fish-pole bamboo, *Phyllostachys aurea*

FIRST, YOU NOTICE a single, slender shoot emerging at the edge of the lawn. You're not sure what it is, but within a week it's been joined by several others. Your neighbour's bamboo is on the run, and it's coming into your garden. What can you do about the very unwelcome intruder from next door?

A Escapee bamboo isn't easy to stop: how much of a nuisance it's going to be depends on whether the original plant is a clumper or a runner. While both types have been known to run away, the latter are much more invasive and can be challenging to get rid of.

Some bamboo species can be astonishingly tough – there are cases of shoots coming up through both asphalt and concrete. As a first step, you can trace them back to source (this may call for some digging) and cut down through the rhizomes coming under the fence with the sharp edge of a spade. If you're unlucky and the bamboo is one of the most invasive, this can end up as a weekly activity; a longer-term, though hard-labour, solution is to sink a vertical barrier – usually of metal, although sometimes of a material not unlike heavyweight vinyl flooring – at least 60cm down, through the soil below the fence. This usually solves the problem, as the rhizomes don't tend to go far downwards, although they can travel a long way horizontally.

If you don't want to go to the length – or rather depth – of sinking a barrier, you're left either with the laborious option of taking shoots out wherever they arise, or of using chemical controls on shoots where they come into your garden.

If the plant is one of the really invasive types, the chemicals will see the shoots off but are unlikely to kill the plant, although if you opt to take this route, read up and take advice on how to use them, and warn your neighbour that that's what you're going to do.

A BAMBOO OF YOUR OWN

What if, rather than trying to contain a neighbour's breakaway bamboo, you'd like to grow one in your own garden? First, take care to pick one of the less invasive varieties. You can grow it in a container, either sunk in the soil, or standing on the ground – although bamboo rarely show at their best in containers, tending to thrive better planted in the ground. If you opt for the latter, consider planting it within a sunken barrier that will contain it in the soil. Specific material is sold to do this job; although it can represent rather a lot of digging, it's worth it. Make sure you take advice from the nursery or garden centre where you buy the bamboo.

There's a huge choice of clump-forming bamboos, and their nature makes them less invasive in the garden. The types to look for are *Bambusa*, *Chusquea*, *Dendrocalamus*, *Drepanostachyum*, *Fargesia*, *Himalayacalamus*, *Schizostachyum*, *Shibataea* and *Thamnocalamus*. If you want to narrow it down further, *Fargesia* and *Shibataea* bamboos are some of the most widely available, and the easiest for a beginner.

The following four are good options:

Umbrella bamboo, *Fargesia murielae*. This species has a height of up to 4m and a spread up to 1.5m. It has yellow-green canes and bright green leaves, with the elegant arching growth that is characteristic of *Fargesia*.

Chinese fountain bamboo, *Fargesia nitida*. This has a height of up to 4m and a spread of up to 1.5m. This *Fargesia* has arched purplish stems and dense foliage.

***Fargesia* Red Panda 'Jiu'.** The 'Jiu' cultivar has a height of up to 4m and a spread of up to 1.5m. It has canes that turn from green to a tawny red in the plant's second or third year, complemented by vivid green foliage.

Ruscus-leaved bamboo, *Shibataea kumasaca*. This has a height of up to 1.5m and a spread of up to 1m. It is a delicate, slow-growing dwarf bamboo, with slim pale-green canes that darken to brown as they mature, and long, deep-green leaves.

Q Can I use ornamentals for structure?

IF YOU'VE TENDED TO THINK paths, walls and hard landscaping are what gives a garden its form, you may not have considered that plants, the garden's *raison d'etre*, are the most important structural element of all.

A

Plants are just as crucial for structure in the garden as any other element – but most will take some time to achieve their full effect after planting. So clever gardeners mix quick growers (which can be removed later) with slow growers (to provide long-term structure).

▼ Clipped pyramids or balls of evergreens can look as effective in a small garden as in a large one.

Clipped evergreens are perhaps the most familiar plants used as punctuation in a garden – for example, clipped balls or shapes of yew acting as permanent features around beds and borders that are mostly planted up with annuals and herbaceous perennials. Narrow, tall conifers are also often used as accents in a similar way. Start to think of plants in terms of their structure and shape, rather than their colour and foliage, and you'll soon see how many options there are for different locations.

FOUR ARCHITECTURAL PLANTS

Here are just a choice few appealing examples – from literally hundreds of highly 'structural' possibilities:

Strawberry tree, *Arbutus unedo.* This has a height of up to 8m after 15 years. A large shrub with a neat, rounded outline and deep green foliage, it carries white flowers and orange-red strawberry-like fruits at the same time, in late summer.

Strawberry tree (fruits)
Arbutus unedo

Cabbage palm, *Cordyline australis.* This tree has a height of up to 8m after 20 years. An evergreen perennial that looks like a palm tree, with a single trunk and clusters of sword-shaped foliage.

Japanese maple, *Acer palmatum* 'Katsura'. This has a height of up to 4.5m after 20 years. A small, elegant and very slow-growing maple with yellow-green, red-bordered leaves, it can be shaped to a clean and architectural outline.

Japanese angelica tree, *Aralia elata.* This has a height of up to 6m. It is a small deciduous tree with huge leaves and white flowers followed by purple berries. Its trunk is so narrow that the foliage can give a fountain effect.

Cabbage palm,
Cordyline australis

Can plants grow between paving?

IF A PAVED COURTYARD or driveway looks a bit barren, what is the best way to 'green' it? Can plants thrive if they're literally planted in the tiny gaps between paving slabs?

If you want to plant between your paving, lift a paving stone before you begin, and look at the soil underneath. It's most likely to have a thin layer of sand directly below it; scrape this away and look at the earth – scratch up a little with a garden fork. If it's friable and crumbles when you rub it between your fingers, plants should 'take' relatively easily. If, however, it's sticky and slick, it indicates a heavy clay soil, which may not work so well.

▼ Creeping thymes, such as this *Thymus serpyllum*, grow easily and are also tough enough to stand up to being walked on without coming to harm.

A

If you pick the right plants, they will happily colonise the gaps in paving. And using them this way is good for wildlife as well as giving a plainly paved area more visual appeal.

Planting up paving

Start by brushing a mixture of compost and sand into the cracks between the paving slabs to give the new plants something to get going on.

Choose which plant varieties you want – there are some ideas in the box, opposite, and buy them either as small plants or as – even smaller – plugs. Divide them gently, making sure that each tiny piece has both root and leaves, and push them into the cracks between the paving stones. Water them, then leave well alone. After a week or two you should notice them starting to settle in and spread. Although most of the varieties that are suggested here will be able to deal with some footfall, you should avoid constantly trampling on the plants, especially while they're settling in.

CHOOSING YOUR PLANTS

The plants that colonise tiny cracks usually have a creeping habit – they will run across the compost, putting down roots as they go. Creeping thymes, such as *Thymus serpyllum* 'Snowdrift' (which, as the name implies, has pretty white flowers), Corsican mint, *Mentha requienii*, and chamomile 'Treneague' are all available as small plants or plugs. *Erigeron* daisies (most commonly Mexican fleabane, *Erigeron karvinskianus*, with its cheerful mixed white and pink flowers, well-loved by bees and butterflies) will also settle happily in tiny spaces. Acaenas are another possibility – *Acaena microphylla* 'Copper Carpet' has deep russet leaves that will create a good contrast with *Erigeron* if you grow them together. Some small sedums, such as *Sedum spathulifolium* 'Purpureum' will also grow in paving in full sun.

Of the above plants, only Corsican mint is also content to grow in shade. If your paving is shaded, other options include red-leaved willow herb (*Epilobium crassum*) – which you may have to raise from seed; this is more easily obtained than small plants – and *Mazus reptans*, which is equally low-growing, with pretty mid-blue flowers.

▼ Planted between paving stones, creeping thymes will tolerate a fair amount of footfall.

Q Where should I site a statue?

YOU CAN CREATE AN APPEALING VISTA even in a small garden; using a statue or sculpture, or even a huge pot, as a focal point often works well. But are there any rules on where in the garden to site it?

A Trying a sculpture out in different spots is an option with smaller pieces, but bear in mind that these generally work best when placed in unexpected places. A larger statue looks good as a focal point, but try placing them off-centre to avoid an overly regimented effect.

If you're still at the stage of choosing a statue, think in terms of matching the piece to the general style of the garden: while you don't always have to site a Buddha statue in a grove of bamboo, say, it might look slightly out of place on a raised plinth in a highly classical terrace. If you want some help in considering what kind of art will work best, print out a few pieces in different styles at a small scale and try cutting them out and superimposing them on a photograph of your garden. It will give you an instant feel for what might work – and what definitely won't.

Think, too, about how a statue or other piece of art will look in different seasons: does it need to be softened by foliage? Will it be best in relatively plain surroundings? Some pieces come into their own in winter, offering something interesting to look at when not much is happening horticulturally.

Larger statues

A larger statue or other object will make a good focal point to catch the eye from a distance. It's often best not to go for bang-in-the-middle obvious when placing it – it may look better slightly off-centre, to avoid an overly regimented effect. Take the time to look at the object from lots of different angles before you settle on its final location: a focal point is there to be walked around, rather than seen from only a single distant spot. Don't forget that some garden furniture, too, can be used to make a strong focal point: a sculptural seat, for instance, can look good from a distance, as well as being functional.

Smaller statues

A smaller piece usually works best when placed where it will come as a visual surprise to people walking round the garden – perhaps on a small stand near to, or even in amongst, a densely planted bed or border. It's fun to place things where people come upon them unexpectedly, as an accent to the atmosphere of the garden, rather than the focal point.

Can my garden have too much colour?

THERE'S NOTHING WRONG with a vividly colourful garden, but sometimes an exclusive concentration on a rainbow of plants can lead to a 'bitty' effect, especially if you have a natural tendency towards the buy-one-of-every-plant-I-like style.

Most gardeners find that if a garden looks somehow unsatisfactory, it's not really a question of too much colour, but of a lack of focus: the colour, rather than uniting, is pulling the eye in different directions. If that's the case, it may be that narrowing down to a defined palette (and thinking of other aspects, such as form) will lead to a more cohesive result.

Which colours?

When you're planning for the next season, fix on a broad palette, such as cool – blue, silver, purple, white – or hot – red, orange, yellow – and plant to reflect that palette.

Follow the traditional gardener's rule of planting in threes or fives or, if you have space, even more – think in terms of blocks or swathes – rather than single examples. This will mean that as the garden comes into flower, you'll have definite areas of colour, rather than the sporadic spots produced by a single, lonely oriental poppy or helenium coming into bloom.

A

Since it's all a question of taste, there's no single answer to this question – unless it's 'it depends'. Many long-term gardeners have gradually moved from the exciting options of maximum colour (and form, and scent...) towards something subtler as they gain gardening experience, and become more appreciative of the finest detail nature offers. But it's a journey that can be taken in the opposite direction, too.

Highly seasonal flowers, such as dahlias (pictured) or tulips, offer you the option of introducing areas of vivid colour for a relatively limited time.

Is decking desirable?

DECKING WAS SO OVERUSED in the 1990s that it became a victim of its own success, and attained the status of a gardening cliché. This is a pity because, used in the right space, decking can be practical, attractive and appropriate. Is it possible to restore its appeal in the garden?

Decking looks at its best in an open, sunny, and ideally raised site. Anywhere smaller, shadier and damper, decking won't work so well. Other aspects of its situation – the house it's alongside and the landscape it's fitting into – will also affect how good decking looks. If it is used near or alongside water, it will automatically look appropriate because if its visual similarity to a jetty or boardwalk.

And a deck will look much more comfortable, too, if it is next to a streamlined modern white or timber-clad house – and less appropriate sitting alongside a classic Victorian or Regency villa. So if you like decking, take a hard look at the situation you're proposing to place it in before deciding for or against.

A

In the right situation, decking can indeed be desirable – just as, in the wrong one, it can become dank and unappealing.

Decking needs to be washed down at least annually if it's not to become very slippery underfoot. A power washer does the best job.

Decking paths

You can, of course, take the 'boardwalk' association literally and make yourself an irregular decking path in your garden. If you have an area where the planting overgrows the path, or you want a raised walkway around a pond, decking can be both an attractive and appropriate option.

THE THREE MAIN TYPES OF DECKING

Hardwood

Hardwood decking is expensive and must be sourced from sustainable woods. It ages to an appealing silvery finish which blends into natural surroundings. Longer-lasting (around 30 years) than other materials, it's the most costly material to fit – it should be fixed with stainless-steel screws to help limit the amount the boards warp as they age. Hardwood can accumulate algae and get slippery if not regularly cleaned; pressure washing at least once a year will remove this problem.

Softwood

Softwood decking is cheaper than hardwood (and is also easier and cheaper to fit), but has a shorter lifespan, although even softwood, carefully maintained and treated with preservative, can be expected to last for around 15 years. Just like hardwood, it will get slimy and slippery if it's not cleaned regularly.

Composite

Composite decking (right) is made from a mixture of wood and plastic. The plastic – either new or recycled – is mixed with wood-dust. The end effect looks convincingly like wood, and composite decking is the cheapest both to buy and to fit. It doesn't get slippery like the two wood options, so requires less maintenance. One downside is that most recycling facilities cannot separate the fused – wood and plastic – elements, meaning that composite decking isn't recyclable.

▼ It's expensive, but hardwood decking may be a worthwhile investment: it's hard-wearing and it greys attractively with age.

Q Why does my garden look boring?

BORING GARDENS TEND TO FALL INTO TWO CATEGORIES: those that have been planned without much imagination, and those that started off looking good but have gradually lost their focus over the years.

A

To work out what the garden needs, ask yourself why you think it looks boring: does it lack a focal point or points? Is it all very symmetrical and even, so that the eye travels over it without catching anything of interest? You may need take time to figure out what's wrong, but when you know, you'll be able to fix it.

Perhaps understandably, you can get so used to certain things that you don't really notice when they stop adding to the appeal or interest of your garden: maybe there's a small tree that's never done well in its situation, or Japanese anemones that used to provide a bright focal point but have spread until there are stragglers all over the garden, or a rose that's an energetic climber but in a rather garish shade of pink.

Planning

The difficult part is to look at your garden with a fresh eye. Start by drawing up a ground plan and fill it in as though you were designing the space from scratch. Use your imagination and pretend that time and money are no object – after all, you don't have to implement the plan; you are just working to give yourself a sense of the possibilities.

Reviewing

Then go back to the real garden. If you don't feel confident of your own judgement, enlist a friend whose taste (and/or garden) you admire. Then make a list of what it would be feasible to do from your plan. This might include things like changing the lawn from a square to a more curved, organic shape, or creating a divider between two areas of the garden to create 'rooms'. At the same time, make a note of things you really don't like, whether they're plants or structures.

At the end of the process, you'll have a list of things to do. Start with a clear-out. Be brave and take bold decisions: rip out anything that isn't earning its keep or has become too big or too crowded. Mature trees are the exception to this rule – cutting down a large tree is not an action that can be undone, so first make sure that you won't change your mind. Then gradually start to implement the changes you've decided on. If it's your first attempt at a major garden overhaul, be prepared to change your mind as you go: the initial clearance is likely to give you different and often more creative ideas of what might look good.

Be prepared to make mistakes

If you're not sure how one of your ideas might look, try it and see. All good gardeners are prepared to make mistakes: one of the joys of gardening is that very little you do is irreversible.

CALL IN THE PROFESSIONALS

And what if, after you've taken a long, hard look at your garden, you still feel utterly stumped? Look up local garden designers, find one whose work you like (there will be examples on their websites) and ask them to comment on your garden for a modest, pre-agreed fee. Most will be happy to do so; there isn't any need for expensive drawings or visualisations – all you want are ideas to act on.

Q Why is my pond level low?

IF A POND that's always been at much the same water level suddenly lowers dramatically, what does it mean? Has it sprung a leak, or is there some other explanation?

A The water levels in ponds do vary season by season to some extent, especially if there's been a long dry spell. If the level keeps falling though, there may be a leak.

Place a tub of water next to the pond and measure the water levels in the pond and the tub. After a week or ten days, measure the water levels again. If the pond level has fallen faster than the level in the tub, there's a leak.

Flexible liner, rigid liner or concrete?

Whatever type of pond you have, wait until the water seems to reach a level and stay there – that is, a point at which it's no longer losing water – then top up the water again (always use rainwater from a butt, not tap water) and wait. If it once again reaches the same level, the likelihood is that there's a leak at this level, and this can usually be repaired. Rubber kits can be used for flexible liners, fibreglass repair for rigid liners, and cement 'wash' for concrete.

To repair any type of pond, you will need to empty it and remove any fish to a large container of water (other wildlife will remove itself as you work). Don't leave the fish in full sun.

Empty and dry the liner before you make the repair, and leave the repair to dry for the amount of time specified in the kit before refilling the pond.

▼ You can test the degree to which a pond's level is falling by comparing it with the level in a bucket of water placed next to the pond.

To repair a concrete pond

Allow the surface to dry after the pond has been emptied, then use a wire brush to abrade the concrete all around the pond from the top to 10cm below the point of the leak. Use a paintbrush to moisten the surface of the abraded concrete, then mix

PROTECTING YOUR POND LIFE

Don't panic about the wildlife in and around the pond. Most kinds of pond life have adapted to living with less water for periods, and it's natural for pond levels to fluctuate quite a bit with the seasons. If, though, you have fish and the water level falls to as low as a 30cm depth, net over the surface while you're considering what to do with the pond; this will protect the fish from becoming vulnerable to predators while the water level is so low.

cement and water to a loose paste and paint it on to the dampened surface. Leave this cement wash to dry, then paint over it with a sealant made especially for concrete ponds (you can buy this at a nursery or online) and leave it to dry thoroughly before refilling the pond.

▼ Most leaks can be mended, although you will almost always have to drain the pond (and evacuate the wildlife!) before you fix them.

If, after being repaired, the water level starts to drop again, chances are that the liner needs replacing or that the concrete may need redoing completely. Sometimes an effective repair can be made to a badly cracked concrete pool by using a liner to cover the whole area; the alternative, if it is badly damaged, is to break up the whole pool and start again: you might opt to replace it, to fill it in, or to turn it into a bog garden.

Q What if I can only use containers?

IN MANY PATIO OR TERRACE GARDENS without open soil, pots and containers may be your only planting option. How can you get the most out of pots and containers and make a satisfactory garden?

Fewer, larger containers will be easier to manage than dozens of small pots, which will dry out quickly and limit the size of the plants you can grow. If you need to be fairly economical, mix up inexpensive containers with more attractive ones. You can even use plastic dustbins with drainage holes bored in them. Mask the less appealing containers by putting the nicer ones in front of them.

A

A container garden has as much potential as any other kind – and is more flexible than some. Look at planning and grouping, creating year-round interest with a wide range of plants or, alternatively, pick a theme that suits your space.

Practicalities

Use heavy, soil-based potting composts and remember that the potting medium will need to be fed monthly and refreshed or topped up once a year or more, as the plants won't continue to do well in tired soil.

Busy lizzie, ***Impatiens walleriana***

Banana palm, ***Musa*** **spp.**

WHICH PLANTS?

To give a container garden a lavish effect – like a 'real' garden that just happens to be in pots – grow a wide variety of plants: some perennials, some annuals, a few herb containers and one or two really sizeable shrubs or even small trees. Alternatively, you can theme by type of plant: in a hot courtyard, tropical choices such as banana plants, phormiums, cacti and succulents will do well; conversely, if your space is naturally shady, you can look at growing the handsome evergreen shrub *Fatsia japonica*, hostas (though guard well against slugs) and a range of ferns.

Alpine wood fern, *Dryopteris wallichiana*

Japanese aralia, *Fatsia japonica*

Consider installing an irrigation system if you have a lot of pots and you don't want to water daily in hotter weather – or if you're away a lot. They aren't expensive and can save a lot of work.

Once the containers are planted up, keep up the maintenance so that they look their best. This means regular dead-heading, checking for (and treating) pests and diseases and replacing plants that don't thrive.

Q How can I make a windy site work?

WHEN A GARDEN IS VERY WINDY, what is the best way to combat it? And are there any plants that will be tough enough to survive – or even thrive – the effects of being blown around?

To work, a windbreak must reduce the effect of the wind without blocking it. The most effective allow 50 to 60 percent of the wind through, cutting its strength.

A

Wind in a garden needs to be filtered – any attempts to block it out will simply create turbulence around the sides of the barriers. To succeed, take a two-pronged attack – first consider growing hedges to act as small windbreaks, and second, look for plants that are up to a blowy challenge and that will grow on, regardless.

Man-made options include vertical lathes or pales, arranged with an amount of space equal to the width of each pale between them, nailed to horizontal supports, and secured to sturdy posts every few yards. In the long term, though, hedges are probably the most attractive and the best way to deal with the effects of wind in a medium-sized garden, and are also great for dividing the garden up into different areas, with the right care (see pp156–159). You will need to keep the area around them clear of other

▼ Unusually for deciduous hedges, beech looks good year-round. Rather than dropping in autumn, the bronze leaves hold on until new growth emerges in spring.

WHICH PLANTS?

There is a wide range of wind-tolerant plants available. Here are five to be recommended.

Oleaster, *Elaeagnus angustifolia* 'Quicksilver'. This has a height of up to 4m. It is a large, good-looking shrub with eye-catching silver leaves, as well as yellow flowers between May and June, followed by yellow fruits.

Grass root, *Eupatorium purpureum*. This deciduous, clump-forming, hardy perennial has a height of up to 2.4m. Althought the foliage is unexceptional, the plant has wonderful, large, showy flowerheads of deep pink between August and October.

Mexican feather grass, *Stipa tenuissima*. This has a height of up to 1m. Many grasses will work well on a windy site, but this deciduous type, although relatively small, is particularly effective, with feathery heads that bow gracefully in the wind.

Common ironweed, *Vernonia fasciculata*. This plant has a height of up to 1.2m. It has tall, tough stems with pointed strap leaves, bearing clusters of fluffy pinkish-purple flowers in summer and is popular with butterflies.

Sea holly, *Eryngium variifolium*. This has a height of up to 40cm. It has spiky silvery-blue flower heads in late summer, growing up from a flat rosette of deep green leaves with much paler veining.

vegetation for at least two years after planting to allow them to establish properly, but the effort will be worth it in the following years, as they'll break up and filter the gusts enough to give plants growing in their shelter a chance.

What kind of hedge to grow depends on the effect you want it to have. Beech (*Fagus*) and hornbeam (*Carpinus*) can look smart if clipped regularly, while hazel (*Corylus*) or field maple (*Acer campestre*) for example, will have a more informal effect.

Any of these options can be planted mixed in with evergreens such as holly (*Ilex*) Portugal laurel (*Prunus lusitanica*) or Ebbinge's silverberry (*Elaeagnus × ebbingei*) to make handsome mixed hedges.

Q Are poultry good for my garden?

IT'S AN APPEALING PICTURE, often seen in lifestyle magazines: chickens pecking happily in open grass alongside glorious flowerbeds. But does this reflect poultry-keeping reality – and will a few free-ranging hens wreck the borders?

Experienced chicken keepers laugh when they see glossy pictures of chickens strutting around beside immaculate herbaceous perennials – they know that the chickens will have been corralled from their usual run to pose for a few minutes only, or the flowers would be looking much the worse for wear. Chickens have large feet, intrusive beaks and a natural scratch-and-peck habit, but there are plenty of ways to give them everything they need while keeping them away from the parts of the garden where they can do damage.

A

Hens can make short work of a plot if they're imported by the unwary, but they can be very good for your garden – by eating pests and producing eggs and manure.

Chickens come in numerous varieties; take advice from an experienced poultry-keeper before choosing which is best for you.

What hens need

How you keep your hens will depend on your circumstances – the ideal is hens in open grass, such as an orchard, behind a gate and with areas they can scratch up for a dust bath, but few gardeners have the luxury of this kind of space.

Hens' most basic requirements are a hen house where they can roost at night (and be shut up to keep them secure from foxes) and space where they can peck, scratch, feed and explore during the day. This can be as simple as a fenced-in permanent run; a temporary run that can be moved around the garden to give different areas a rest, and give the hens a change of scene; or a closed-off portion of the garden – behind a barrier or gate – where the hens have what they need, but

THE PROS AND CONS OF POULTRY

Pros

- They eat a lot of pests, and will even eat slugs.
- If you rake up hens' droppings and compost them along with your other garden waste (they burn plants if applied directly), these make a useful manure.
- Enthusiasts find hens endlessly entertaining to watch and say they have a lot of personality.
- The eggs – of course. Invariably these will be fresher than any you can buy, and eating them will feel much more satisfying.

Make sure your birds are properly protected from foxes, as these predators can strike at any time.

Cons

- You need to be vigilant about foxes. They can visit during the day or at night, so chickens must be properly protected.
- Parts of the garden will need to be shut off – particularly in spring when hens can inflict terminal damage on tender young plants.
- Hens need looking out for: chickens, like any other animals, can suffer from a range of pests and diseases – inform yourself so you'll be able to spot problems.

without access to your more tender or treasured plants.

Whichever sort of house-and-space combination you opt for, you can feel reassured that your hens will be having a happier life than their commercially raised cousins. Bantams are always a possible alternative to larger chicken breeds – smaller all round, their feet make less impact in the garden, although of course the eggs they lay are smaller, too.

Can I have a roof garden?

IF THE ONLY OUTSIDE SPACE that you have is a roof, you're bound to wonder if it could be made into the garden that you don't have at ground level. But what are the practicalities of making a roof garden, and what do you need to check on before you get started?

First, get a structural survey done by a professional surveyor. This is essential: it will ensure that the roof can bear the additional weight of pots, soil, plants, seating – and people – and also that it's watertight. Then consider large containers with evergreens, supplemented by perennials, making sure they can withstand the wind. Keep it tidy and free of dead leaves and flowers.

Once a professional structural survey has established that your roof is up to the job, plan out the different elements of the roof garden. Before you go shopping, remember to think about the following components.

Pots

Large containers are best from a practical point of view: they won't dry out as quickly as smaller ones. Once filled, they will be very heavy, so look at the possibilities of plastic or fibreglass containers that look like terracotta or lead: there are plenty of attractive fakes.

▼ Keeping a range of herbs in securely fastened pots in your roof garden means you can pick home-grown supplies as you cook.

Watering

Do you need a watering system, or are you happy to water daily? If you don't have an irrigation system, think about where the water is to come from: you may not want to carry it up from the nearest tap two floors down, so it does need thinking through.

Windbreaks

Carefully chosen plants can help here, but roofs are windy spots, and you may need a non-plant windbreak, too, if you want to sit out. Light fencing or woven hurdles, or canvas strung across a frame can all do the job, but will need to be anchored securely.

Plants

A 'backbone' of evergreens is a good idea, and you can supplement them with perennials. Plants that don't need a lot of water will obviously be at an advantage and if yours is a city rooftop, it's likely that it will be frost-free unless in exceptionally cold weather: think about succulents, sedums, agaves and yuccas. A herb garden will be an appealing addition: group them in a large, low container, where you can pick leaves in passing.

Seating

Tubular metal chairs or tables are a good choice for rooftops: they're relatively light, but not light enough to blow away if they're put in a sheltered corner. Very light plastic or very heavy metal options are best avoided.

NOT QUITE A ROOF GARDEN...

... but increasingly popular are green roofs for sheds or outdoor office buildings. They look pretty, can be applied to even tiny buildings, and are (relatively) easy to make provided that you have average DIY skills. You do need to check that the building you're thinking of 'greening' can take the extra weight; home-constructed green roofs are essentially frames containing compost which is planted up, so they add quite substantial heft to an average shed roof. Of course professional versions are also available, for a price. At a time when every green metre should be valued, a green roof, however modest, will also offer an extra haven for wildlife.

▼ Sedums, here tasteless stonecrop (*Sedum sexangulare*), thrive in the generous supply of light available on a roof, and are appealing to wildlife.

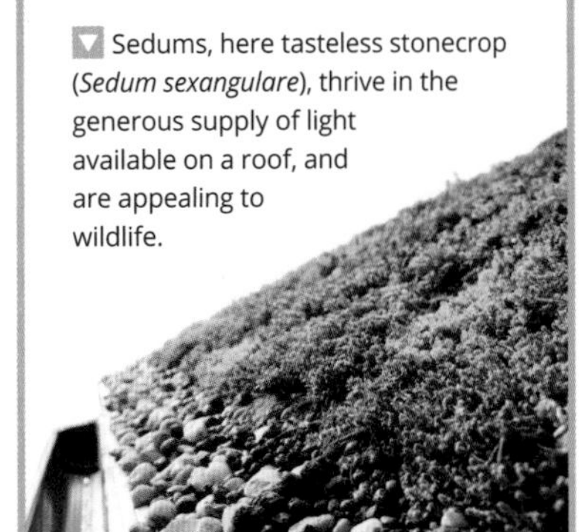

Q Where should I sit outside?

IT'S A COMMON MISTAKE when a gardener comes fresh to a new space: there's a seating area somewhere obvious – directly outside the back door or in a corner of the rather small lawn – so they put out chairs and a table, without ever thinking whether it's the best area for sitting. What makes outside seating areas work and where's the best place for one?

A Even keen gardeners usually spend plenty of time outside relaxing as well as tending to the garden. Rather than plonking down chairs and a table where they've always been, look at the different factors – sun, shelter, outlook – that will make the best lounging spot.

First, consider when during the day different areas of your garden get the sun. There's no point deciding on a seating area that catches the midday sun with just the right amount of shade across one corner if you're never at home in the middle of the day. If you like to take a cup of coffee outside in the early morning, or sit in the evening sun when you get in from work, look at where you'll get the best light when you can use it. If the problem may be too much light rather than too little, consider a pergola with climbers trained across it to create a green roof that will offer attractive dappled shade.

More than one

Even if your garden is small, you don't need to limit yourself to a single seating area – there's usually space for a couple of chairs and a coffee table in one corner and a bench in another. Build seating areas into the overall prospect of your garden: it's actually nicest if the garden encroaches a little, rather than being too rigorously cleared around a sitting-out spot. If you want to 'signal' seating space, consider laying a small area of brick or paving to delineate it, and then plant around it to soften the edges.

Use your imagination to create different kinds of seating area around a garden. Even a small plot usually has space for more than one.

TOP CLIMBERS FOR OVERHEAD GREENERY

If you want something scented to climb up a pergola, these five, from a wide range of options, will all give you fragrance as well as shade. With the exception of *Clematis armandii*, which prefers full sun, all are sun-lovers but can cope with at least partial shade.

***Rosa* 'Madame Alfred Carrière'.** This is a vigorous, deliciously scented climber with large, creamy-white flowers and almost thornless stems.

***Rosa* 'Danse du Feu'**. If you want vivid colour as well as fragrance, this cultivar has bright scarlet, strongly scented flowers from midsummer to early autumn.

Italian honeysuckle, *Lonicera* x *italica*. This honeysuckle is an energetic climber with particularly large and strongly scented flowers, which open white and gradually turn to a honey-yellow.

Poet's jasmine, *Jasminum officinale*. The classic 'common' climbing jasmine, with its highly characteristic sweet scent and pretty white flowers, is unfussy and easy to grow.

Armand clematis, *Clematis armandii*. Ideal if you have lots of space and a large pergola, this evergreen plant has dense, glossy leaves and large, white, scented flowers from early to mid-spring.

Poet's jasmine, *Jasminum officinale*

Right plants for the right space

Add scented plants nearby, making a choice depending on the time of day that you're most likely to be sitting out: scented stocks and nicotianas are great for evening fragrance, while scented geraniums and lavender will give good value during the day.

Brompton stock, *Matthiola incana*

Q Will trees take over?

IF SPACE IS LIMITED in your small garden, how realistic is it to plant trees, or at least a tree – and if it's feasible to have just one, how can you to narrow down the selection?

A

Most spaces can cope with a small tree, or at least a large shrub. Make sure you see a grown example before making your final selection – and don't forget to measure the available space.

Opinion on what the height range of a 'small' tree should be varies. Professionals usually set it at between 8 and 12 metres; amateurs sometimes feel that this definition of 'small' is much taller than they envisaged.

Thinking ahead

The most important thing to do before you go tree shopping is to measure accurately the amount of space you have available for a tree in your garden.

If you plot the garden shape onto paper, you can super-impose the spread of any tree you're interested in, and that will give you a good idea of how much of the garden it will take up; spread is probably more important than height when it comes to how the tree will look *in situ*. In a very confined space, a columnar shape will work better than a tree that is bushier. Check, too, how fast a tree will grow – growing speeds vary a lot and if you want (relatively) fast results, something like a maple or cherry tree can put on as much as 60cm a year (sometimes even more). Professional gardeners sometimes plant a fast-grower for effect and a slow-grower nearby, getting rid of the first after 10 years or so, and letting the second take over, although this is probably too ruthless a tactic for most amateurs.

While it's far more unusual than most people think for trees to damage building foundations – it happens only very occasionally and then only on shrinkable heavy clay soils – it's sensible to avoid planting very thirsty types, such as willows, too near the house. If you are in any doubt about planting a particular tree in a certain spot in your garden, get expert advice from a chartered surveyor first.

When you have a realistic idea of how much space you have (and how much light your tree will have), narrow down the options by considering specifics: what do you most want – spring blossom? Richly coloured autumn foliage? A magnet for wildlife? Or maybe neighbouring gardens are

FIVE GOOD DECIDUOUS TREES FOR A SMALLER GARDEN

Remember to consider the spread as well as the height of a tree. 'Fastigiate' trees – trees of a naturally columnar, upright shape – can be particularly useful choices for smaller spaces.

Paperbark maple, *Acer griseum*. This has peeling reddish-brown bark and bright orange and red foliage in autumn. It has a height of 8–12m and a spread of 4–8m.

Broad-leaved cockspur thorn, *Crataegus persimilis* 'Prunifolia'. This has glossy deep-green leaves that turn red and orange in autumn, and pretty creamy-white hawthorn-like blossom. It has a height of 4–8m and a spread of 8m and over.

Mountain ash, *Sorbus* 'Joseph Rock'. This has an upright shape and white flowers followed by striking yellow berries. It has a height of 9m and a spread of 7m.

Showy crab apple, *Malus* x *floribunda* (above). This has narrow leaves and an upright shape, with white or pale pink flowers and small red-and-yellow fruit. It has a height of 8–12m and a spread of 8m and over.

Cherry, *Prunus* 'Amanogawa'. This is a column-shaped cherry with delicate pink flowers and bronze foliage in spring. It has a height of 8m and a spread of 3m.

already full of cherry or magnolia trees, and you'd like your tree to be different, or flower earlier (or later)? Ideally if you have space for only one tree, it should offer interest in more than one season, so that it pays its way with, for example, flowers in spring and colourful bark year-round. When you've filled in your list of practicalities and must-haves, you can look for the dream tree that ticks all the boxes.

Further reading

An Ear to the Ground: Garden Science for Ordinary Mortals
Ken Thompson
Transworld Publishers, 2003

Alan Titchmarsh How to Garden: Climbers and Wall Shrubs
Alan Titchmarsh
BBC Books, 2010

Alan Titchmarsh How to Garden: Pruning and Training
Alan Titchmarsh
BBC Books, 2009

Bob's Basics: Composting
Bob Flowerdew
Kyle Books, 2010

***Charles Dowding's Vegetable Garden Diary: No Dig, Healthy Soil, Fewer Weeds* (2nd edition)**
Charles Dowding
No Dig Garden, 2017

The Complete Gardener: A Practical, Imaginative Guide to Every Aspect of Gardening
Monty Don
Dorling Kindersley, 2009

Dream Plants for the Natural Garden
Piet Oudolf and Henk Gerritsen
Frances Lincoln, 2013

Drought-Resistant Gardening: Lessons from Beth Chatto's Gravel Garden
Beth Chatto and Steven Wooster
Frances Lincoln, 2000

RHS Encyclopedia of Gardening
Christopher Brickell
Dorling Kindersley, 2012

The Fruit Tree Handbook
Ben Pike
UIT Cambridge, 2011

Garden Design: A Book of Ideas
Heidi Howcroft and Marianne Majerus
Mitchell Beazley, 2015

Gardeners' World: The Veg Grower's Almanac, Month by Month Planning and Planting
Martyn Cox
BBC Books, 2014

Good Soil: Manure, Compost and Nourishment for Your Garden
Tina Råman
Frances Lincoln, 2017

RHS Gardening School
Simon Akeroy and Dr Ross Bayton
Mitchell Beazley, 2018

RHS Grow for Flavour
James Wong
Mitchell Beazley, 2015

The Hillier Manual of Trees and Shrubs
John G. Hillier and Roy Lancaster (Eds.)
Royal Horticultural Society, 2014

RHS How Do Worms Work?
Guy Barter
Mitchell Beazley, 2016

How to Create an Eco Garden: The Practical Guide to Greener, Planet-Friendly Gardening
John Walker
Aquamarine, 2011

RHS Lessons from Great Gardeners: Forty Gardening Icons and What They Teach Us
Matthew Biggs
Mitchell Beazley, 2015

Life in the Soil: A Guide for Naturalists and Gardeners
James B. Nardi
University of Chicago Press, 2007

RHS Pests & Diseases: The Definitive Guide to Prevention and Treatment **(2nd edition)**
Pippa Greenwood and Andrew Halstead
Dorling Kindersley, 2018

Planting: A New Perspective
Piet Oudolf and Noel Kingsbury
Timber Press, 2013

The Sceptical Gardener
Ken Thompson
Icon Books, 2016

Collins Tree Guide
Owen Johnson
Collins, 2006

Index

Credits

Cover, 3, 84 (top), 97 (aphid, flea beetle), 154 (top), 162 © Morphart Creation | Shutterstock. 4, 7 (top), 218 © Hein Nouwens | Shutterstock. 5 © MoreVector | Shutterstock. 6, 219 © geraria | Shutterstock. 7 (bottom) © Bodor Tivadar | Shutterstock. 8 © Paustius | Shutterstock. 9 © pandapaw | Shutterstock. 11, 41, 76, 95 (left), 107 (left), 108, 176 (top) © RHS, Lindley Library. 13 (top), 34, 85 (top), 89 © Richard Griffin | Shutterstock. 13 (bottom) © Kuttelvaserova Stuchelova | Shutterstock. 14 © Stocksnapper | Shutterstock. 17 © dokoupilova | Shutterstock. 18 © Julia Karo | Shutterstock. 19 © Andrii Spy_k | Shutterstock. 20 © Niagara705 | Shutterstock. 21 (from top) © RHS / Leigh Hunt, © RHS / Carol Sheppard, © Dominicus Johannes Bergsma | Wiki Commons, © T.Kiya | Wiki Commons, © Garden World Images Ltd / Alamy Stock Photo, © T.Kiya | Wiki Commons. 22 © Julius Elias | Getty Images. 23 © Olga Gorevan | Shutterstock. 24 © By Leah-Anne Thompson | Shutterstock. 25 © Nattika | Shutterstock. 26 © oksana2010 | Shutterstock. 27 © Andrea Jones Images / Alamy Stock Photo. 29 © Fir Mamat / Alamy Stock Photo. 31 (bottom) © Kisialiou Yury | Shutterstock. 32 © ubonwanu | Shutterstock. 33 © Silver Spiral Arts | Shutterstock. 35, 97 (photo, top right) © Imageman | Shutterstock. 35 © Heinz Hauser / botanikfoto / Alamy Stock Photo. 38 © V J Matthew | Shutterstock. 40 © Ivonne Wierink | Shutterstock. 42, 56 (bottom), 139 (bottom) © Scisetti Alfio | Shutterstock. 44 © Andreja Donko | Shutterstock. 45 (bottom), 50, 62, 191 © picturepartners | Shutterstock. 46, 86, 142 (bottom) © RHS. 48, 141 (left), 144 © RHS / Tim Sandall. 51 © kelifamily | Shutterstock. 53 (bottom), 141 (right), 165 (bottom) © Potapov Alexander | Shutterstock. 55 © RHS / Wendy Wesley. 56 (top) © Stephen B. Goodwin | Shutterstock. 60 © Anjo Kan | Shutterstock. 63 (from top) © Gita Kulinitch Studio | Shutterstock, © kongsky | Shutterstock, © Cameramannz | Shutterstock. 64 © claire norman | Shutterstock. 65 © Melica | Shutterstock. 66 © Neirfy | Shutterstock. 68 © Alena Brozova | Shutterstock. 69, 77, 121, 139 (top), 145 (right) © Madlen | Shutterstock. 70 (left) © Nik Merkulov | Shutterstock. 70 (right) © JIANG HONGYAN | Shutterstock. 71 © Anton-Burakov | Shutterstock. 75 © Mark D Bailey | Shutterstock. 78, 142 (top) © RHS / Pathology. 79 © Garden World Images Ltd / Alamy Stock Photo. 81 © Graham Corney | Shutterstock. 82 © Binh Thanh Bui | Shutterstock. 84 (bottom) © LAURA_VN | Shutterstock. 85 (bottom), 204 © Sarah Marchant | Shutterstock. 87 © S and O Mathews Photography / Alamy Stock Photo. 90 (top) © irin-k | Shutterstock. 90 (bottom) © Zzvet | Shutterstock. 91 (left) © Ezume Images | Shutterstock. 91 (right) © TierneyMJ | Shutterstock. 93 © Marbury | Shutterstock. 94 © schankz | Shutterstock. 95 (right) © Peter Andrus | Wiki Commons. 97 photos: (top left) © akepong srichaichana | Shutterstock, (middle left) © Alexander Raths | Shutterstock, (middle right) © Dionisvera | Shutterstock, (bottom left) © Superheang168 | Shutterstock, (bottom right) © Lepas | Shutterstock. 99 © MICROSONE | Shutterstock. 100 © videnko | Shutterstock. 101 © Matt Pagett. 104 © IgorGolovniov | Shutterstock. 105 (top) © Mark Herreid | Shutterstock. 105 (bottom) © Steve Heap | Shutterstock. 106 © oksana2010| Shutterstock. 109 © AS Food studio | Shutterstock. 110 © IrinaK | Shutterstock. 111 (top) © Antonio S | Shutterstock. 111 (bottom) © sasimoto | Shutterstock. 112 © Lindsey Johns. 113 (top) © DUSAN ZIDAR | Shutterstock. 113 (bottom) © aussiegall | Creative Commons CC-BY-SA 2.0. 114 © Drew Rawcliffe | Shutterstock. 115 © MarjanCermelj | Shutterstock. 116 © Menna | Shutterstock. 118 © Captiva55 | Shutterstock. 120 © nikkytok | Shutterstock. 122 (top) © Carpe Diem - Flora / Alamy Stock Photo. 122 (bottom) © Dylan OHara | Shutterstock. 124 © Wellcome Collection. 125 (top) © kzww | Shutterstock. 125 (bottom) © S. Rae | Creative Commons CC-BY-SA 2.0. 126 (top) © kamnuan | Shutterstock. 126 (bottom) © JUN3 | Shutterstock. 127 (top) © Tom Biegalski | Shutterstock. 127 (bottom) © KariDesign | Shutterstock. 128 © xpixel | Shutterstock. 129 © RHS / Rachael Tanner. 130 © allotment boy 1 / Alamy Stock Photo. 131 © Carlos Delgado | Creative Commons CC-BY-SA 3.0. 133 © D. Kucharski K. Kucharska | Shutterstock. 134 © Wiert nieuman | Shutterstock. 135 © Ariene Studio | Shutterstock. 136 © David Lee | Shutterstock. 137 (top) © Studioimagen73 | Shutterstock. 137 (bottom) © Sarah2 | Shutterstock. 138 © On_Ter | Shutterstock. 143 © focal point | Shutterstock. 145 (left) © Smileus | Shutterstock. 148 © gualtiero boffi | Shutterstock. 149 © Chutima Kee | Shutterstock. 152 © dabjola | Shutterstock. 153 © Leigh Prather | Shutterstock. 154 (bottom) © N-sky | Shutterstock. 156 © sabza | Shutterstock. 157 © Tim Gainey / Alamy Stock Photo. 158 © IreneuszB | Shutterstock. 159 © John-Kelly | Shutterstock. 160 © steamroller_blues | Shutterstock. 163 (clockwise from top) © Eric Isselee | Shutterstock, © Pixel Memoirs | Shutterstock, © irin-k | Shutterstock, © DioGen | Shutterstock. 164 (top) © Andrew M. Allport | Shutterstock. 164 (bottom) © mccw thissen | Shutterstock. 165 (top) © Erni | Shutterstock. 166 © cosma | Shutterstock. 167 (bottom) © Tainar | Shutterstock. 168 (left) © RHS / Beatrice Henricot. 169 © ansem | Shutterstock. 171 (clockwise from top) © dabjola | Shutterstock, © Jan Holm | Shutterstock, © kavring | Shutterstock, © Angorius | Shutterstock, © DK Arts | Shutterstock. 172 © Ivaschenko Roman | Shutterstock. 173 (bottom) © Serenko Natalia | Shutterstock. 175 (left) © LiliGraphie | Shutterstock. 175 (right) © prajuab lekpetch | Shutterstock. 176 (bottom) © IraFrank | Shutterstock. 177 (top) © Mikhail Melnikov | Shutterstock. 177 (bottom) © gailhampshire | Creative Commons CC-BY-SA 2.0 . 178 © Silo | Shutterstock. 179 © Alexander Raths | Shutterstock. 180 (top) © Anest | Shutterstock. 180 (bottom) © RHS / John Glover. 181 (left) © Andrey_Kuzmin | Shutterstock. 181 (right) © BookyBuggy | Shutterstock. 183 © lynea | Shutterstock. 185 © Beautiful landscape | Shutterstock. 187 the808 | Shutterstock. 188 © davemhuntphotography | Shutterstock. 189 © GAP Photos. 190 © Del Boy | Shutterstock. 192 © DK Arts | Shutterstock. 194 © GAP Photos / Elke Borkowski . 195 (top) © Angel Simon | Shutterstock. 195 (bottom) © Vahan Abrahamyan | Shutterstock. 197 © GAP Photos / Elke Borkowski . 199 © Hong Vo | Shutterstock. 200 © bubutu | Shutterstock. 201 (top) © Ben Schonewille | Shutterstock. 201 (bottom) © iMoved Studio | Shutterstock. 202 © Simone Voigt | Shutterstock. 205 © Jeanie333 | Shutterstock. 206 (left) © Praneet Soontronront | Shutterstock. 206 (right) © TunedIn by Westend61 | Shutterstock. 207 (top) © GAP Photos/Fiona Lea. 207 (bottom) © Vitaly Raduntsev | Shutterstock. 208 © Manuo | Shutterstock. 210 © Tsekhmister | Shutterstock. 211 (top) © Eric Isselee | Shutterstock. 211 (bottom) © NinaM | Shutterstock. 212 © Franz Peter Rudolf | Shutterstock. 213 © josefkubes | Shutterstock. 214 © ballykdy | Shutterstock. 215 (top) Tamara Kulikova | Shutterstock. 215 (bottom) © natu | Shutterstock.